DUALITY INNOVATION

Engineered by Perfect Imagination

Typeset in Georgia by Sanjeev Kumar Singh

ISBN-10: 1979930015

ISBN-13: 9781979930017

Introduction

Do you exercise and play with your mind at least once per day? If you are a busy person like most people, you may not get around to it every day. But your mind deserves a happier, healthier, longer life, and daily mind exercise is critical to achieving that. Innovation requires a lot of mind exercise. Meet the world's first gravity defying innovation technique for innovators that launches a new way of thinking when inventing. In just minutes, after making a couple of derivations, connections and relationships, your mind can come up with one or more solutions to even most intractable problems in the technical world! And it's fun for both you and your career.

A duality-driven designing is committed to changing an inventor's relationship with their pet project through innovative product designing that promotes a unique and dominant solution. Developed by an IP industry leader, this book offers something perfect for every life stage of a product development cycle.

Sanjeev K. Singh, Ph.D.

Alpharetta, GA, USA
December 2017

CONTENTS

CHAPTER No. | PAGE No.

Chapter 1

DUALITY: THE VERY FOUNDATION OF OUR UNIVERSE

DUALITY controls everything, I mean EVERYTHING. It is a law of nature that exists everywhere there is a physical world in the universe. If this is the FACT, then why should we ignore it. Knowing that even a stopped analog watch shows correct time twice a day tells us how pervasive DUALITY is. While ONE thing could mean TWO different things or TWO different things can mean the same ONE thing. This is the magic of duality. Nobody can beat it so it is best to respect it. Agreed.

To understand duality imagine looking at your hand and recognizing there is a "FRONT" and there is a "BACK" of your hand both of which always coexist in such a way that when one is present the other is present by default. No force or power or energy can separate the existence to these two realities -- the "FRONT" and the "BACK." In other words, the front cannot exist without the back and vice versa. Once one grasps this simple understanding then you can visualize that they both are equal to each other yet being opposite in time and space. The real-world consequence of this realization is that if you have one view point about a subject then there is an equal possibility that there can be an opposite viewpoint of that subject by default due to dual nature of things. If it is always true that at least two viewpoints will exist, you would be foolish to believe that your viewpoint is the ultimate or absolute truth.

In order to incorporate the duality in our daily lives, routine situations or relationships, the first step is to become open-minded enough so you can at least hear out the perspective of the other side. This way you can understand their rationale and once understood will have less urge to fight it. The golden rule of the duality is to seek balance as the "FRONT" and "BACK" of the hand exercise during their time of existence. They respect each other's personal space and coexist simultaneously without fights or disturbances.

Mind it, the concept of the duality is not a new one as the realm of its effect has been known for thousands of years but what is indeed new is full recognition of it having unforgiving omnipresence in every sphere of our life so pervasive that one cannot escape its influence even in hell, if there is such a thing. There is no need to turn away from such a powerful force but it is best to embrace it and honor it for being the ultimate source of creation – live or lifeless.

When we have clarity that my side of the story is not the only side of that story we can feel free to engage with others who may have quite opposite angle on the same facts or happenings. Clearly one does not have to persist on own truths but learn to appreciate truths of others with whom you may have differences on thorny issues. This type of growth mindset can help resolve even the most trickiest of situations. And one can handle any intractable problem with a level of generosity and mutual acceptance of existence of deviating thought processes. A shared cause of understanding each other albeit wide gaps should be the goal in view of duality.

The first step in a resolution of a dispute is dialogue and being curious to hear out the other person. There is also a need to be not very picky in defining things rather staying flexible so views of both parties can be accommodated in the final settlement or solution. By listening to other side with an open mind does not mean you will not make your voice heard. On the other hand, a healthy relationship is the one in which both individuals win and are equally heard.

Duality reinforces this notion of double possibilities or two approaches. Sometimes we need to settle on one of many options. Duality is not a zero-sum game where if one wins then other one loses. It is an ongoing dialogue where one wins on certain occasions and the other wins on some other occasions in a way that both people feel like not looser. Basically, both persons have to believe in this dual nature of things and be fair and generous to each other so that peace is maintained in the relationship and in terms of power balance the relationship is not lop-sided for a girl or a boy and a man or a woman.

Chapter 2

DRIVE INNOVATION: BY APPLYING THE DUALITY PRINCIPLE

NO innovation is easy. Some can happen with brute force or following time-tested methodologies. But wouldn't it be great if there in an ANOTHER WAY or method to THIS madness. Agreed.

In the world of "it is WHAT it is" there is no innovation that can happen. Innovation is hard work and a lot of sweat. Some innovation can happen by brute force method by exhausting many possible possibilities. Some innovation can happen by applying standard innovation techniques known in engineering circles. But wouldn't it be nice if one can have a magic wand and apply a recipe to cook up solutions to even very difficult technical problems. I have always wondered can the DUALITY principle be applied in problem solving.

We know that duality exist everywhere so the question is can we find a way to exploit this knowledge for innovation purposes. For example, in one field where duality can be applied is a "DESIGN AROUND" situation. My thought is that since most innovation happens in increments, there is always a last best-known solution exists in engineering. So, a new design project can become a "DESIGN AROUND" project. You have a start point and you have an end point to get to.

EXAMPLE I

Here is an example. I invented a 2-connection DOG seatbelt harness. The known design at the time included a vest for the torso of the dog and a harness to tether the dog to a buckle of a vehicle at a car seat where there is a seatbelt present. The problem with this solution was that in an accident the dog will be thrown away based on the length of the harness as the buckle connected acted as a pivot point and the harness acted as a leash with certain length. See FIG. 1 below.

ALL of the Dog Seatbelt Harnesses in the Market are of 1 Connection Point Design

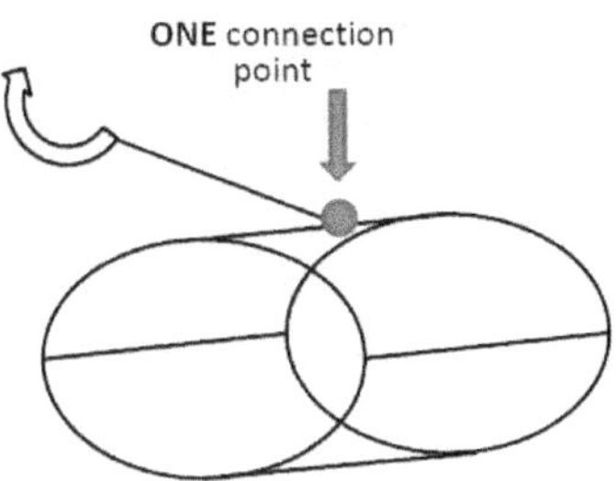

By using the DUALITY principle, I added a second connection or a harness to the vest and floated the dog in between the buckle and the prong so the seatbelt webbing retracting system can be tapped to protect the dog just like the humans. See FIG. 2 below.

DUALITY here meant going from one connection to two connections. So, duality in different inventions would mean different things depending upon specific circumstances. That's the hard part of using this concept. I wish there was a straight-forward simple formula to apply and get the benefit of this approach. However, my thought is with practice one can see how to use some opposite effect than what people have already used. It could be as easy as just the opposite logic.

My **New Improved Patented Design** of 2 Connection Point Dog Seatbelt Harness

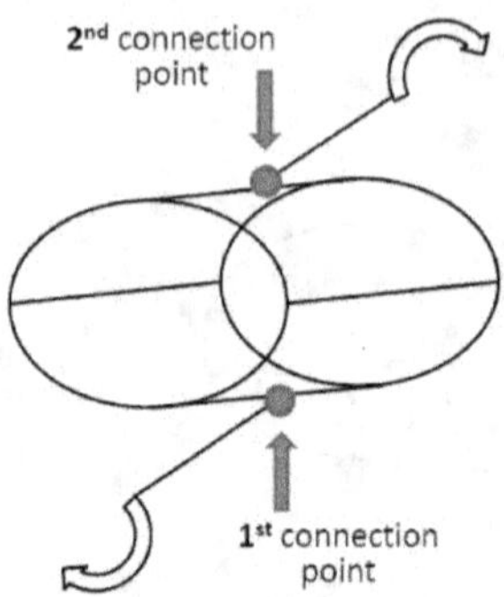

Benefits of 2 Connection Point over the 1 Connection Point Seatbelt Harness

Safety and Comfort FEATUREs	ONE Connection Point Design	2 Connection Point Design
Lock Dog's Position while allowing Dog to stand and move	✗	✓
Limit how far Dog can go away from the car seat	✗	✓
Allow Dog to stand and move easily and comfortably	✗	✓
Provide differentiation from all the other products to sell/market	✗	✓

How to apply the DUALITY principle and invent?

How to think about this invention in terms of duality innovation. The problem here is to tether the dog such that it is not a pivot from a point. In other words, the dog is at the end of one connection. To avoid the pivot effect, you can put the dog between two connections. The first connection is connection "1" and the second connection is connection "0" which is flexible as it can expand and retract. So, think how you can use duality in a given problem setup. We know the existing design used a single connection and had serious problems. The leap here is to deploy something more to fix the issues. Then the question is what more. Recall the seatbelt sits on the belly of a human being and the person gets anchored with belly in the middle. The two portions of the webbing on top of the belly and the bottom of the belly can be seen as two connections keeping the belly safe in the middle while the seatbelt stretches and shrinks over the belly. Imagine the belly of the person as a dog and fit it between two opposite connections. So one connection is fixed "1" and one flexible "0".

EXAMPLE II

Here is another example. I was designing a solution to prevent a license tag sticker posted on a vehicle's license plate in most states of the U.S. So, I came up with a mini-frame of the size of the license tag sticker which overlapped the edges of the license tag sticker to protect from being peeled off. Consider this as solution called number “0” or “empty.” See FIG. 3 below.

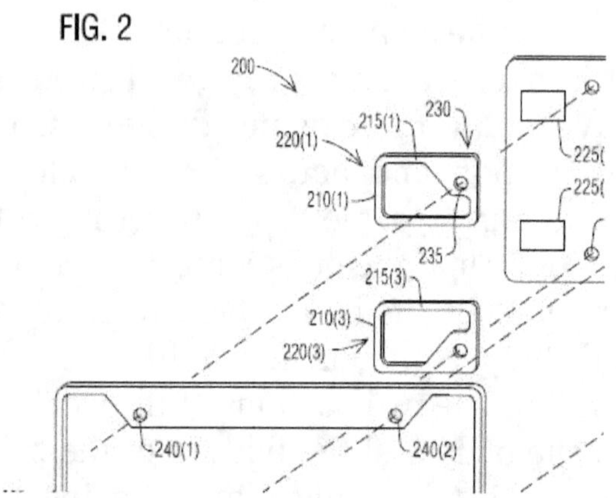

Then, I came up with a DUALITY solution as number “1” or “full.” The number “1” or “full” represents a transparent plate covering just the license tag sticker. See FIG. 4 below.

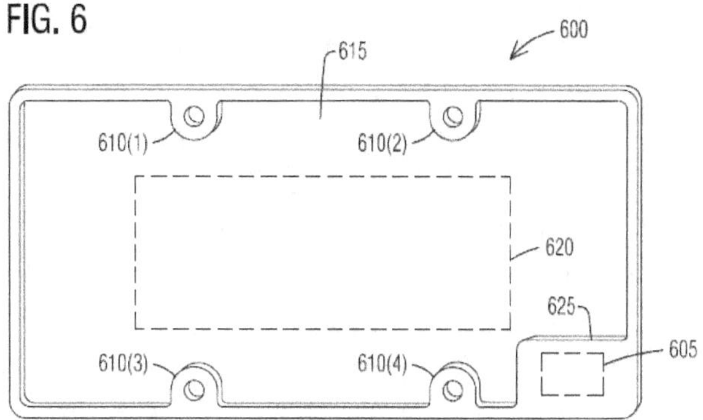

How to apply the DUALITY principle and invent?

How to think about this invention in terms of duality innovation. The problem here is to cover the license tag sticker. You can apply duality and say you can cover the edges by a mini-frame as "0". Or you can completely cover it with a transparent layer and call it "1". There are only these two possibilities which you can discover by the above application of the duality principle.

EXAMPLE III

Here is another example. A game console with two wireless accessories docked on it.

Let us imagine the game console to have a display screen in the middle of the device as

"1". Then, on two sides – left and right, there can be wireless accessories that can be docked. The idea was to have a compact design of the game console where two wireless accessories can be retained with a mother device such as the display screen in the center. These two wireless accessories can be "0'" and "0"". The benefit of this configuration is that the wireless accessories can be charged when docked. And there is a less chance of that the two wireless accessories can be lost as they have a place to keep them with the display screen of the game console. See FIG. 11 below.

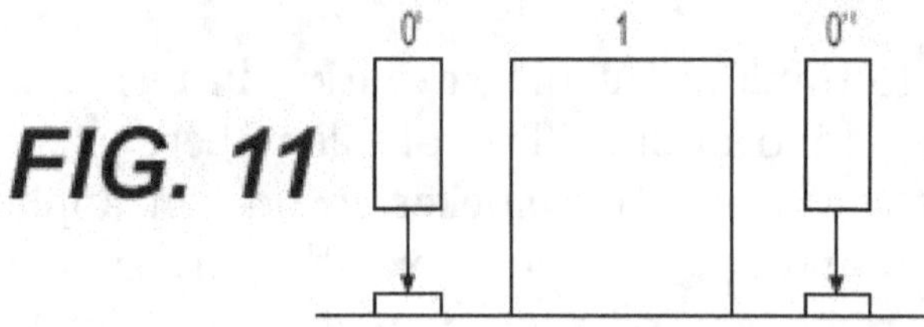

How to apply the DUALITY principle and invent?

How to think about this invention in terms of duality innovation. The mother device is "1" so we can make a combination with a component/part as "0". This "0" can be anything which has some reason to be with the mother device "1'. I thought of the "0" as wireless accessory which can be docked at the mother device “1”. So, you start with the duality and fit your solution to this format and can invent new device features or new devices etc.

Chapter 3

DETAILED ANALYSIS OF RELATIONSHIP OF DUALITY TO MY 15 INVENTIONS

I have always wondered can the DUALITY principle be applied in problem solving. We know that duality exist everywhere so the question is can we find a way to exploit this knowledge for innovation purposes. Agreed.

1. Baby Adhesive Bandage

A typical adhesive bandage includes two arms extending from a medicated pad region in the center. These 2 arms can be considered “0”. To make this typical adhesive bandage choke proof for babies, two much longer arms were added perpendicular to the original 2 arms. We can call these 2 new arms as “1”. So, a combination of two opposite “0” and “1” instances provided a final solution. See FIG. 1 below.

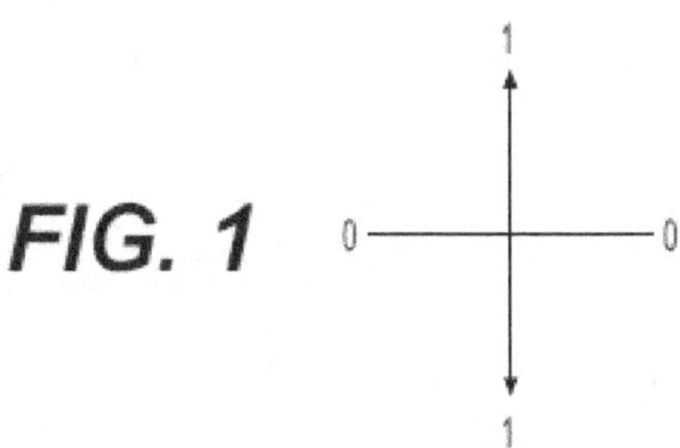

2. Two-wheel drive bicycle

In a typical rear-wheel drive bicycle, a drive mechanism is installed with a sprocket and a paddle. We can call this "1" – () – "0" system being on a left side of the bicycle. To build a two-wheel drive bicycle, a second drive mechanism is provided on other side of a bicycle frame as a second "1" – () – "0" system. See FIG. 2 below.

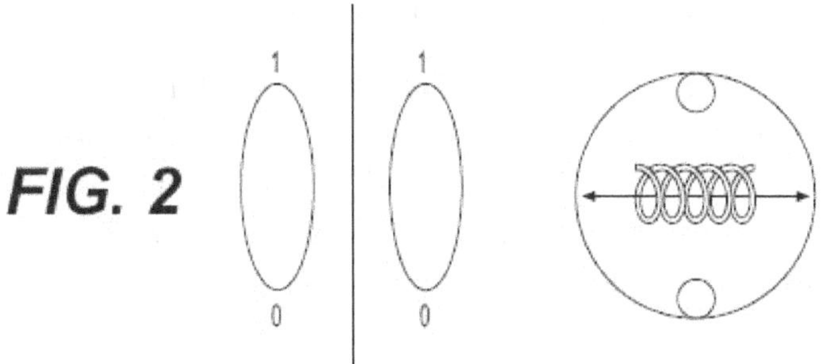

This combination provides a weight-balanced bicycle. Of course, a deslacker device was needed, which includes a spring set in two opposing tubes having a gap being mounted on a drive chain installed between two sprockets on extreme ends. The spring in this configuration provides expansion force to maintain a constant slack-less drive mechanism. Here slack can be "1" and the opposite can be spring "0" to counter the slack.

3. An audio-visual audio book

In one case, a user interface will have two regions "1" and "0" – one region for displaying actual hard-copy or eBook book images including figures, graphs etc. and other region for providing various audio controls for a user playing the audio book in the user interface of an audio book APP.

In another case, with one region only the top area will be "1" and bottom area would be "0".

The top area will display the actual book images and the bottom area would display audio controls. The "1" and "0" representing two different forms of information that go hand in hand as the audio-visual audio book.

See FIG. 3 below.

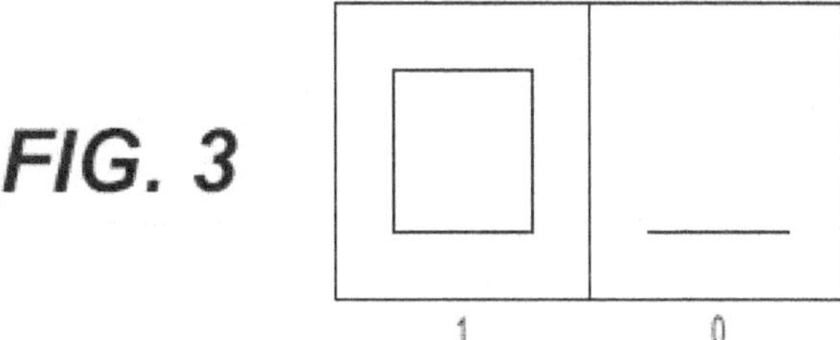

4. Breakable 6-pack band for 6 cans

In the old or known design, a tab was provided to peel it such that when the tab was peeled, a ring was broken and a CAN gets released. However, the CAN be easily removed – the way it has been historically done – without operating the tab. So, most people will just ignore the tab and just pull the CAN from the 6-ring band by brute force. See FIG. 4 below.

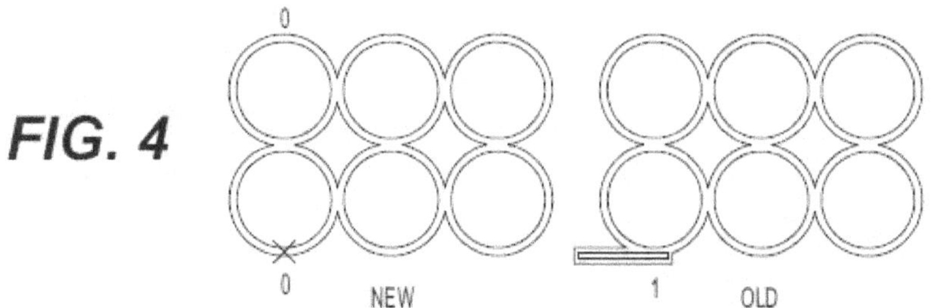

Knowing this shortcoming of the current solution “0”, I figured we needed that the CAN can only be removed if the ring is broken 100%, the solution “1”. In other words, the ring breaking is NOT optional but mandatory. To make this happen, a glue dot is used to glue the 6-ring band to the neck of the CAN.

5. Hotel soap bar

The current hotel soap bar design is generally a reason for waste of soap because the bar is solid and it is used only few times typically before it is discarded. This is design “1”. We need the design “0” – a hollow soap bar. See FIG. 5 below.

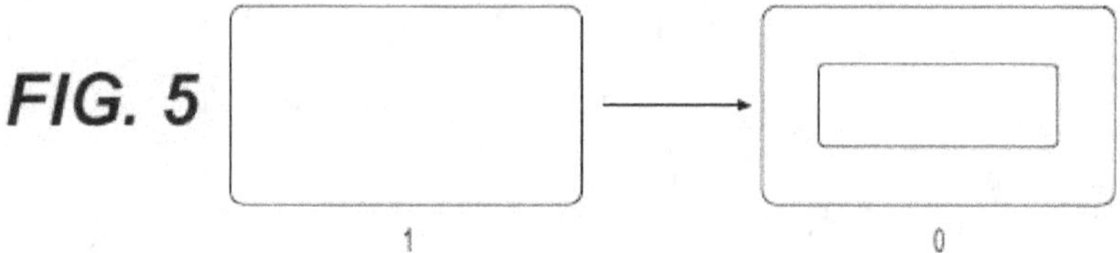

This is not just a hollow soap bar but a soap bar which is built using an inner box of another material such as paper so the manufacturing is a cheap and doable process. For example, a molten soap process is used to manufacture the hotel soap bar. So, the invention is a combination of materials of choice and a process of manufacturing being a choice as it is very cost effective to make a cheap product in an environment friendly way.

6. Smart wristband

This is a classic case of use of DUALITY in design. Arrival of a smart watch has meant for many people letting going wearing their analog or mechanical watches. A smart watch has brains in a watch case. If you want to have an analog or mechanical watch in the watch case you can transfer brains from the watch case “1” to a watch band “0”. See FIG. 6 below.

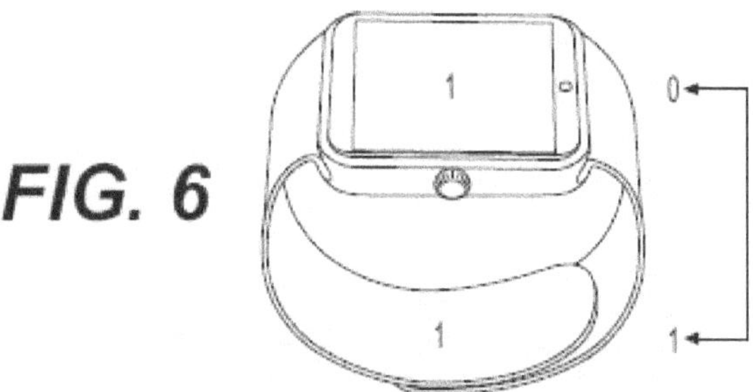

In this way, you end up with a SMART wristband with all the brains of a smart watch built in the wristband itself. So, it is about transferring brains to the opposite logic so you can use your choice of watch case – not a smart one but an analog or mechanical one while you have the smart watch functionality built in the wrist band. Sort of reversal of the logic between a case and a band.

7. Car occupancy alarm

Thirty plus kids die every year in hot cars due to heat strokes when they are left locked in the vehicles by mistake. The challenge was to build a device which will sense a situation and communicate in the event of a problem. The sensing part was easy – a voice and motion sensor. To know the car is parked, an ignition off sensor was needed. But the problem was how to communicate with a parent's cell phone without a WI-FI or a cellular signal. See FIG. 7 below.

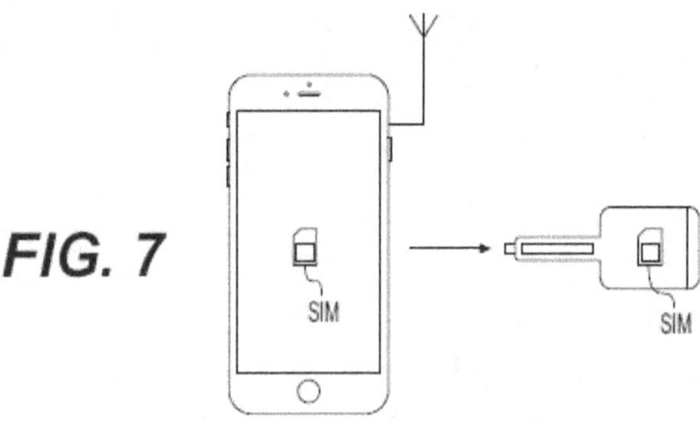

This was solved by making a mapping in a cell phone feature and a new device. Cell phones come with SIM cards to make a phone call. The new device was provided a SIM card just like the new APPLE watch has, an ability to self-call based on sensor data. Here the DUALITY was about doing the mapping from one type of device to a new needed device.

8. Phone disabler in a vehicle

There are thousands of deaths that occur due to distracted drivers. Mobile Phone is generally the culprit which entices drivers in texting or browsing, leading to mishaps endangering themselves and others on the road. There are many solutions to avoid this outcome. But none is in background and beyond the user control. Also, the solution has to figure out if one is in a driver's seat or in a passenger's seat. A driver is "0" and the passenger is "1". See FIG. 8 below.

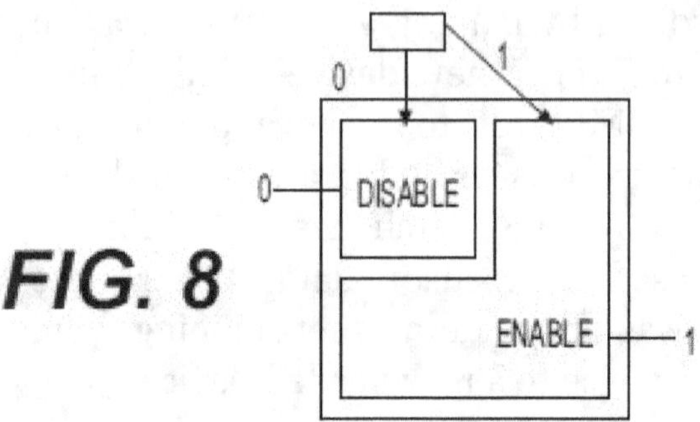

FIG. 8

A Bluetooth signal from a Bluetooth radio of a vehicle is measured by a mobile phone and determine based on location if the mobile phone is of a driver's or a passenger's. The distance to the driver is "0" and the distance to the passenger is "1".

9. Tetherable Booster Seat

The current booster seat is exactly what it says – a seat boosted few inches and that's all. What I noticed that during hard stops the booster seats move forward and endanger life of kids. This design is "0". The best solution is to provide tethers at back corners. The tethers can be anchored at vehicle seat buckles. But if we use a prong in the tethers it will occupy the buckle and it won't be available to connect a prong of a vehicle seatbelt. See FIG. 9 below.

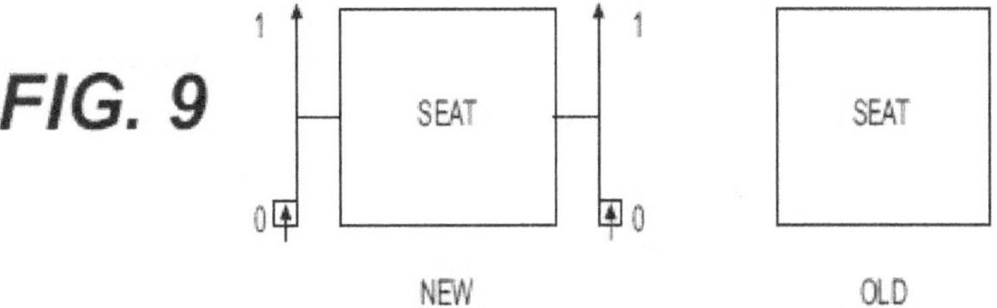

FIG. 9

So, what is needed was a design which can accommodate both the prongs. So, a "1" and "0" pair is attached on both sides of the booster seat. This new feature allows the booster seat to be installed on a vehicle seat on a rear seat being in the middle of 3 seats easily and securely such that the booster seat does not slip when the vehicle is stopped suddenly.

10. Document holder

In this invention, the concept of DUALITY comes into action by viewing two connectors as one pair of opposite functions. In the document holder design there are two connectors – one on the top and one on the bottom. If the bottom one is considered as a "0" connector by being used to hold as a base on to an edge of a laptop screen, then the top connector may be thought as its opposite as being used in for loose papers for temporarily hold them during a typing session. So, this connector can be termed as "1".

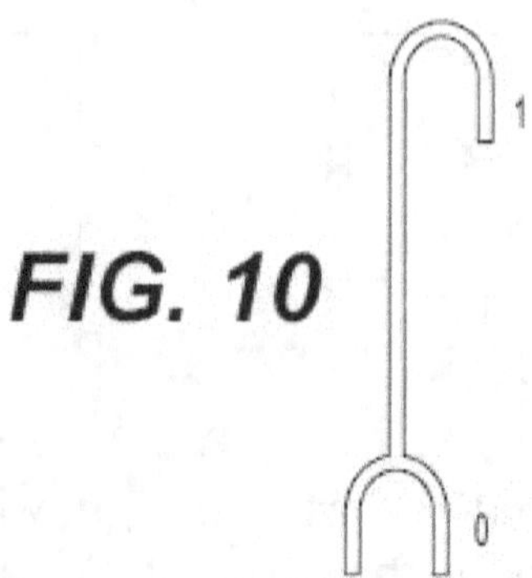

FIG. 10

This distinction becomes clear in the FIG. 10 above, where the below connector has always the laptop display screen edge and the upper connector has paper sometimes.

11. Game console with two wireless accessories

Let us imagine the game console to have a display screen in the middle of the device as "1". Then, on two sides – left and right, there can be wireless accessories that can be docked. The idea was to have a compact design of the game console where two wireless accessories can be retained with a mother device such as the display screen in the center. These two wireless accessories can be "0'" and "0''". The befit of this configuration that the wireless accessories can be charged when docked. And there is a less chance of that the two wireless accessories can be lost as then have a place to keep them with the display screen of the game console. See FIG. 11 below.

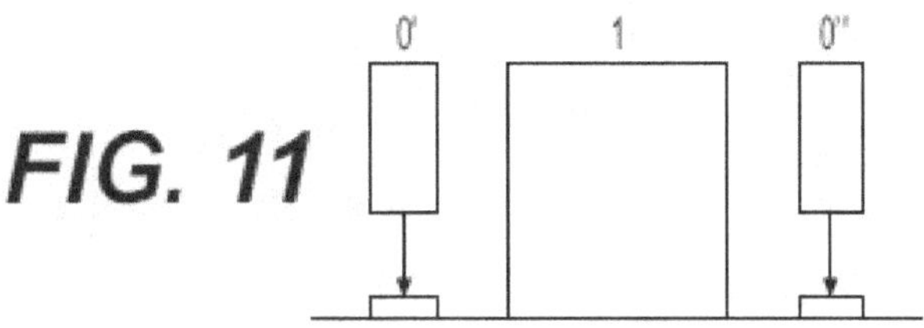

12. Small or big ear Earbuds

Here we have a product designed specifically for people with two extreme size of ear canals. Either they are too small so normal earbuds don't fit and fall out when they jog or they are too big that they are not snugly fit and by being loosely fit come out of the ears in any sports activity. A new design of earbuds was needed which can be of any size between 1 to 10 and work well meaning without above known problems. An earbud with fingers as petals of a flower was designed to fulfill the above set forth specification. The fingers were made of a resilient material so they can expand and shrink, giving different sizes that fit small or big ear canals. On the ends of fingers teeth were provide to lock their size for repetitive use. A closed state on the left in FIG. below is "0" and an open state on the right in FIG. 12 below is "1".

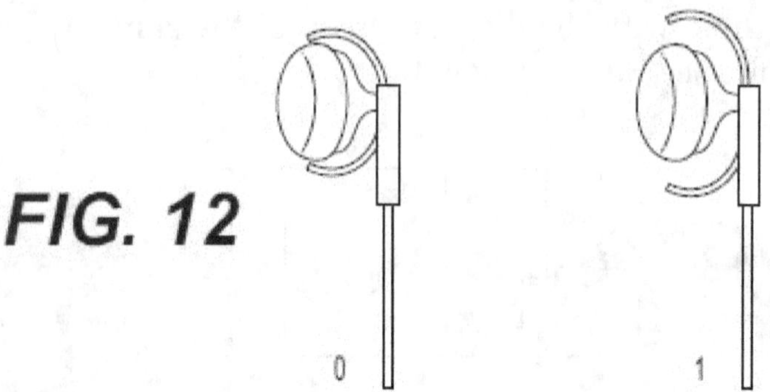

13. Multi illumination state brake light

When I saw that brake lights of luxury vehicles had multiple LEDs instead of an old-style bulb, I thought we can use the LEDs to form a light pattern representative of the level of braking done by a driver of a vehicle ahead. Connecting above dots in which a light pattern could be as simple as a pattern of signal bars on a cell phone where a small bar means little braking and many bars including a biggest bar means hard braking. For 100s of years since a vehicle was born, a "0" and "1" logic has been used for no brake or brake. Now we use logic "1", "2", "3", "4", "5". So old logic can be one digit or bit "0" and new logic can be many or multiple-bit or digit as "1". See FIG. 13 below.

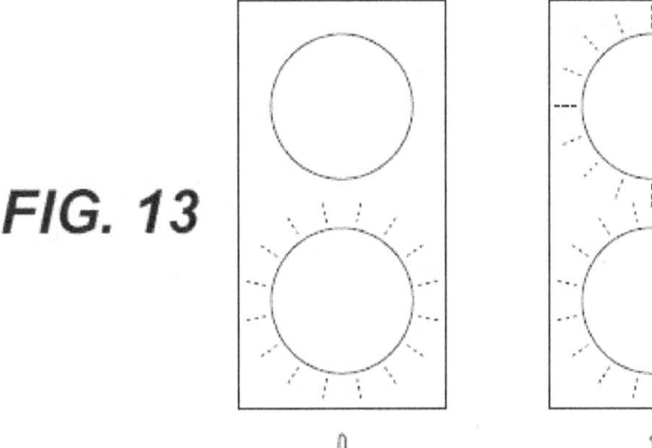

14. 2-connection DOG seatbelt harness

We were on a Texas highway when we spotted a gentle man in a pick-up truck driving with his dog in his lap unrestrained. A search to solve this problem ensued. This led to a 1-connection design of dog seatbelt harness which was of no use in terms of saving the dog's life in an accident. I realized that the 1-connection dog seatbelt harness was missing a second connection to the vehicle seatbelt release and retract system. In this way, 2-connection dog seatbelt harness was born. Let's call the old 1-connection dog seatbelt harness "0" and the new 2-connection dog seatbelt harness "1". One has just one connection and the other has 2 connections. See FIG. 14 below.

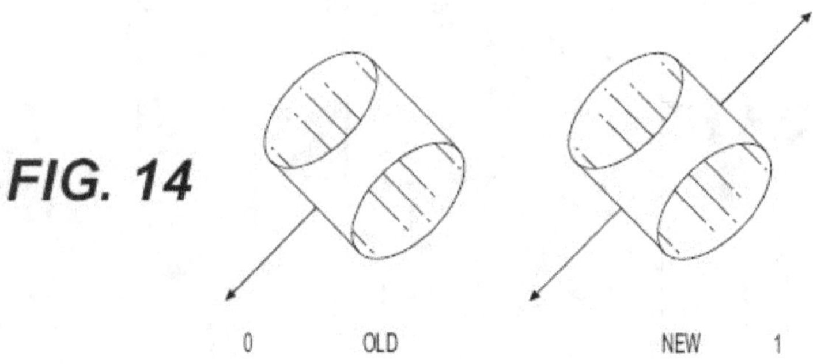

15. License tag sticker protector

In some neighborhoods, your license tag stickers can be stolen on a higher rate than others. But this crime happens. People use the theft to provide registration to no-registered vehicles. A mini-frame with a hole in a corner which is just the size to overlap a standard license tag sticker of any state in U.S. is provided as design “0”. This is an empty frame that just covers the edges of the license tag sticker. A design “1” can be a transparent sheet of plastic that covers fully just the area of the license tag sticker. This is a full design opposite of the empty design “0”. DUALITY can be thought of in such ways and inspire one to find a solution. See FIG. 15 below.

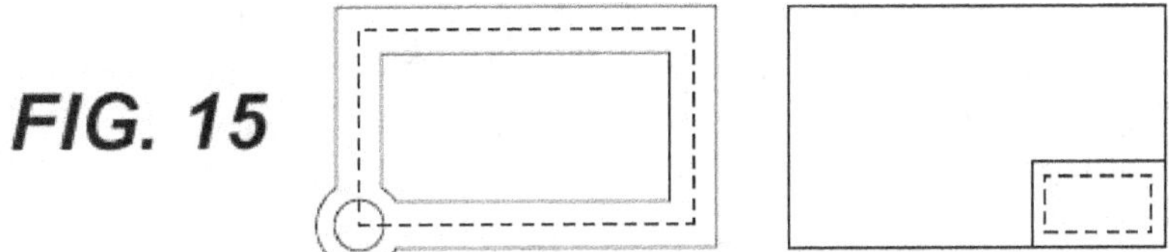
FIG. 15

Chapter 4

2 OF MY PATENTED PRODUCTS LICENSED AND 2 PATENTS SOLD AND ON ONE FILED AN INFRINGEMENT LAWSUIT

We need better products. Agreed

FIRST PRODUCT LICENSED

I have licensed my DOG seatbelt harness to a pet product company. It is an opportunity for the company to explore helping dogs by making their travel safer and comfortable than what is possible today. This improvement in safety and comfort of dogs is accomplished by a new patented design of the dog seatbelt harness. The new design is called "2-Connection Point" dog seatbelt harness. All currently available dog seatbelt harness designs in the market are "1-Connection Point" designs. There are four distinct benefits of the new design over the all the rest of them in terms of safety and comfort to the dog.

A 2-Connection Point dog seatbelt harness is provided which can be worn by a dog on his torso and can be connected to the vehicle seatbelt system at two places, i.e., the prong and buckle associated with a vehicle seat. The 2-Connection Point dog seatbelt harness includes a vest and two harnesses coupled to it. One harness has a vehicle seatbelt buckle connected to its free end and the other harness has a vehicle seatbelt prong connected to its free end. The vest is to be put on the body of the dog and the buckle and prong are to be connected to the buckle and prong available next to a vehicle seat. This way, a dog is connected to the vehicle seatbelt webbing lock and release system and can be protected in hard braking or accident just like a vehicle passenger is protected by the vehicle seatbelt system.

Key Features:

1. Locks Dog Position by using the seatbelt locking system of the vehicle (no product on the market does this and this is a key safety improvement).

2. Limits how far Dog can go away from the car seat so dog cannot injure himself or the passenger in case of the accident.

3. Allows Dog to stand and move easily and comfortably (no product on the market allows dog to stand and move comfortably while tied in the car).

So far I have received 4 U.S. patents on my new design. To know more about the new design see YouTube video.

PROTOTYPE VIDEO: See a video of the prototype of the k9 Seatbelt™ at:

http://www.youtube.com/watch?v=MQK-SSX1ZGM&

My 2-connection dog seatbelt harness design is crash-tested in a Lab for a 30lb dog at 30 miles/hour speed.

SECOND PRODUCT LICENSED

My second product was licensed also. It will save innocent lives in cars of drivers from unsafe texting/browsing while driving. Phone Feature Lock-out APP product designed for a mobile phone (only APP that wirelessly & autonomously disables/enables phone features in a vehicle) (U.S. Patent app. no. 15/234,204 filed)

Lock-out mechanisms for driver handheld mobile devices are disclosed. The lock-out mechanisms disable the ability of a handheld mobile device to perform certain functions, such as texting, while one is driving.

A handheld mobile device position locator and temporarily disabler vehicle system comprising:

a motion analyzer configured to detect whether a vehicle or a handheld mobile device is in motion beyond a predetermined threshold level;

a mobile device position locator configured to determine whether the handheld mobile device is located within a safe operating area of the vehicle based on a wireless Bluetooth-based system, a WLAN-based system, or a Bluetooth-WLAN combo based system having at least one of metric such as a Radio Signal Strength Indicator (RSSI), a received bit error rate (BER) and a cellular signal quality (CSQ) associated with the handheld mobile device; and

a lock-out mechanism configured to automatically and selectively disable one or more functions of the handheld mobile device based on outputs from the motion analyzer and the mobile device position locator.

The motion analyzer may be located in a vehicle and/or a handheld mobile device.

The mobile device position locator is located in the vehicle.

The lock-out mechanism is located in the handheld mobile device.

The mobile device position locator includes one or more Bluetooth stations and/or WLAN stations, which may be access points configured to communicate with the handheld mobile device to determine a location of it in the vehicle in terms of being in a safe area or in an unsafe area. The mobile device position locator measures a Radio Signal Strength Indicator (RSSI), a received bit error rate (BER) and a cellular signal quality (CSQ) associated with the handheld mobile device. These parameters: a Radio Signal Strength Indicator (RSSI), a received bit error rate (BER) and a cellular signal quality (CSQ) associated with the handheld mobile device are mapped to a distance from the driver's seat of the handheld mobile device. For example, a 0.5 meter or 2-3 feet away from the driver's seat is considered safe as the handheld mobile device is determined to be in the safe area. The lock-out mechanism enables the ability of a handheld mobile device to perform certain functions, such as texting or browsing, while one is not driving or not in the driver's seat being in safe area. The lock-out mechanism disables the ability of a handheld mobile device to perform certain functions, such as texting or browsing, while one is driving or in the driver's seat being in unsafe area.

MY 2 U.S. PATENTS SOLD

Business Opportunity

Automotive brake lighting system will fit in with auto business very well since it is unique in a number of ways. My patented system can be incorporated by vehicle manufactures in their vehicles.

Executive Summary

A binary operation of complete turn-off and turn-on of a conventional brake light is unable to provide a good pre-warning effect to a driver of a trailing vehicle. According to my patent an automotive brake lighting system controls powering of light sources of a vehicle brake light to clearly signal drivers of following vehicles a sudden braking and degree of deceleration of the vehicle.

Benefits:

> This system differentiates between the sudden braking and slow braking by energizing more or less light sources of the brake light based on the level of braking.
>
> A sequential switching on and off of light sources provides greater perception of the application of brakes and thus safety.

It alerts the trailing driver of amount of braking forces applied and improves the driver's awareness and reaction time.

By enabling the driver of a following vehicle to have a better idea of a braking rate of the leading vehicle the following driver can act according to the actual braking situation due to an advanced warning of a progressive and variable nature.

Market

In 2009 in US more than 10 million new autos were sold. Source: Autodata as appeared in the USA Today article of January 6, 2010.

Need

Insufficient spacing between vehicles is the cause of one of the most costly and serious types of traffic accidents, the rear end collision. Rough estimates conservatively indicate that it accounts for more that one-third of all traffic accidents and one-half of the resulting injuries. Driving too close to a leading vehicle is the most basic driving error. Tailgating produces the deadliest damage, ranking the number one cause of auto accidents and injuries on the highway. Not only do such accidents result in loss of life and injuries which causes human suffering to persons and their families, but such accidents often require expensive repairs and cause insurance premiums to increase.

Rear end collisions can often be prevented. Generally, such collisions are the direct result of drivers trailing so closely or tailgating, that they cannot stop in time. When the forward vehicle stops quickly or slows down, the driver of the trailing vehicle often fails to appreciate that stopping his vehicle involves numerous factors: e.g., the driver's reflexes, reaction time, age, eyesight, physical condition, awareness of other vehicles, attentiveness and concentration on driving, sobriety, as well as the velocity of the vehicle, condition and grade of the road, weather conditions, visibility, type and weight of the vehicle, operating condition of the vehicle, and especially its tires and brakes. Conditions also vary between daytime and night-time driving and the amount of ambient light available. It is often difficult to quickly determine the minimum safe distance required between cars because of the numerous factors involved.

One safety organization recommends one car length for each 10 miles per hour of speed of the trailing vehicle. Improving a driver's awareness and reaction time is important. For example, providing an improvement of awareness or reaction time of even 1/10th of a second not previously available gives an additional three feet of stopping distance at 20 miles per hour, six feet at 40 miles per hour, and as much as nine feet at 60 miles per hour. This extra distance can be the difference between a safe stop and a rear end collision. Even 10ths of seconds of improvement in awareness or reaction time is of considerable significance.

Significantly, motor vehicle accidents involving rear end collisions comprise about one-third of the total accidents in the United States. Indeed, the severity of this problem prompted the U.S. Government to mandate the use of high-mounted, third brake lights on all post 1985 U.S. delivered automobiles. Through various tests, it has been found that such central high-mounted brake lights provide greater and improved recognition factors to the driver of the trailing vehicle and improves the reaction time of the driver by about 1/10th of a second in a braking condition, i.e. where the leading vehicle has activated its brakes.

Solution

In essence, my patented system (U.S. Patent Nos. 7,755,474 & 7,982,594 entitled Automotive Brake Lighting) will allow a driver to estimate the level of braking of the vehicle in front since the harder the leading driver pushes on the brake pedal, the greater and distinct the light warning to the following driver. Such a variable light warning would provide greater and improved conscious recognition to the driver of the trailing vehicle and improve the reaction time, avoiding costly and serious types of traffic accidents, the rear end collision.

ON ONE OF MY U.S. PATENTS FILED AN INFRINGEMENT LAWSUIT

Consumer devices are all about user experience and exclusive features whether it is hardware or software. One can license my U.S. patents to enhance the user experience by offering two exclusive features based on my 2 U.S. patents.

My 2 U.S. patents (U.S. Patent No. 8,472,658 entitled hand-held, portable electronic device with retainer port for receiving retainable wireless accessory for use therewith) may be used in the wireless game controller of PS3, PSP and PS Vita and even with cell phones to improve user experience by offering exclusive features. On the wireless game controller two USB ports may be provided where the two wireless Bluetooth Headsets can be docketed for retention and in situ charging when the battery of the wireless game controller is charged. This ability will allow the users to always have wireless headsets charged and available at a known location, making remembering to charge and remembering their location redundant. Thus, increasing convenience of use for the users which won't be available in competitor's devices.

A federal lawsuit against GameStop Inc. – a retailer of Nintendo, Inc.'s SWITCH product (a game console) has been file in 2017.

Chapter 5

WHAT ABOUT DOGS

Don't DOGS need protection while traveling in a vehicle just like humans do? YES, they being our furry friends do too. Agreed.

My this invention (U.S. Patent NO. 8,622,431) relates generally to seatbelt harnesses for pets such as dogs and they are intended to be used for their protection during travel in a motor vehicle by directly connecting to a vehicle seatbelt system including a seatbelt webbing lock-release system.

BACKGROUND OF PROBLEM

About 70 million plus homes in United States of America have a pet and 50% of those have two pets. Whether it's a quick trip to the supermarket or a long ride to the beach, companion animals are now traveling animals too and pet friendly lodging has increased 300% since 2005. For example, 82% of pets travel on vacation with their owners. However, 98% of dogs do not travel properly restrained in a moving vehicle. When driving at 35 mph speed, a 60-pound unrestrained dog can cause an impact of 2,700 lbs, slamming into a motor vehicle seat, windshield, or driver/passenger(s). In fact, driver distraction causes more accidents than any other issue.

An average of seventy five percent family pets visit veterinary clinic every year for treatment. Forty percent of all vet fees come from unforeseen illnesses or accidents. Only 20% of family pets are covered by a pet insurance. The average vet bill for attention and care needed following a road accident involving a cat or dog last year was around $1000 to $1,500.

Traveling in a vehicle with a dog can pose a serious danger to the pet. In an accident, a pet--like a person--can exert a force of 20 times its body weight if it is not properly restrained. Should another passenger collide with a pet, serious, life-threatening injuries can result to both occupants. Even worse, a dog can be thrown from a vehicle in a collision. For pet safety during their travel in a vehicle, use of a variety of pet containers or harnesses types and designs is known in the prior art. Several of these known pet containers or harnesses comprise familiar, expected and obvious structural configurations, notwithstanding the myriad of designs encompassed by the many of such pet containers or harnesses which have been devised to fulfill numerous objectives and requirements associated with pet travel in a vehicle.

While these solutions fulfill their respective, particular objectives and requirements, many known pet restraining container or harness systems fail to disclose a way for easier use in a vehicle for safely securing a pet. More particularly, such known solutions do not enable a safe, secure and comfortable environment for a pet traveling in a vehicle as is otherwise available to human occupants of the vehicle. When the pet is traveling within the vehicle, these known solutions to the pet safety during their travel in a vehicle fail to safely and controllably restrain a pet during a collision or sudden breaking of the vehicle.

SOLUTION

Seatbelt harnesses are provided for pets such as dogs for use with a vehicle seatbelt system to restrain them for protection during their travel in a vehicle. For protection of a dog during travel in a motor vehicle, in one embodiment, a seatbelt harness may include a restraint portion coupled to an attachment portion that may have first and second connectors which are either fixedly or removably attached to the restraint portion. The first connector may be adapted to attach to a seatbelt or a first belt connector in their default positions and the second connector may be adapted to attach to a second belt connector or an anchor. The positions of first and second connectors may be selected for aligning them with the restraint portion such that the attachment portion to act in conjunction with the restraint portion to enable positioning of the pet on a vehicle seat at an intermediate point along a longitudinal path between the first belt connector and the second belt connector or between the first belt connector and the anchor.

In one exemplary embodiment of the present invention, a seatbelt harness is provided to restrain a pet such as a dog in a motor vehicle. The vehicle may include a vehicle seat and an associated shoulder lap belt combination seatbelt system that has at least one of a seatbelt connected to a seatbelt webbing lock-release system, a first belt connector connected to the seatbelt, a second belt connector and an anchor that connects to an auto child seat all of which are being provided near or at the vehicle seat. The seatbelt harness may include a restraint portion configured to be worn by the pet on their torso. The seatbelt harness may further include an attachment portion that is coupled or couples to the restraint portion. The attachment portion may have at least one of first and second connectors each of which have a free end. The first connector may be adapted to couple to at least one of the seatbelt or the first belt connector in default positions thereof during a state of non-operating mode. The first connector may connect the attachment portion directly to the seatbelt webbing lock-release system via the seatbelt. The second connector may be adapted to couple to at least one of the second belt connector or the anchor. While the first connector may be configured and arranged at a first predetermined position, the second connector may be configured and arranged at a second predetermined position relative to the first determined position. The first and second predetermined positions may be selected for aligning the first and second connectors with the restraint portion such that the attachment portion to act in conjunction with the restraint portion. This arrangement or configuration may enable positioning of the pet on the vehicle seat at an intermediate point along a longitudinal path between the first belt connector and the second belt connector or between the first belt connector and the anchor.

In another exemplary embodiment of the present invention, a kit for a seatbelt harness is provided to restrain a pet such as a dog in a motor vehicle. The vehicle may include a vehicle seat and an associated shoulder/lap belt combination seatbelt system that has at least one of a seatbelt connected to a seatbelt webbing lock-release system, a first belt connector connected to the seatbelt, a second belt connector and an anchor that connects to an auto child seat all of which are being provided near or at the vehicle seat. The seatbelt harness kit may include a restraint portion that may be configured to be worn by the pet on their torso. The seatbelt harness kit may further include an attachment portion that may be configured to be removably coupled to the restraint portion. The attachment portion may have at least one of first and second connectors each of which have a free end. The first connector may be adapted to couple to at least one of the seatbelt or the first belt connector in default positions thereof during a state of non-operating mode. The first connector may connect the attachment portion directly to the seatbelt webbing lock-release system via the seatbelt. The second connector may be adapted to couple to at least one of the second belt connector or the anchor. While the first connector may be configured and arranged at a first predetermined position, the second connector may be configured and arranged at a second predetermined position relative to the first determined position. The first and second predetermined positions may be selected for aligning the first and second connectors with the restraint portion such that the attachment portion to act in conjunction with the restraint portion. This arrangement/configuration may enable positioning of the pet on the vehicle seat at an intermediate point along a longitudinal path between the first belt connector and the second belt connector or between the first belt connector and the anchor.

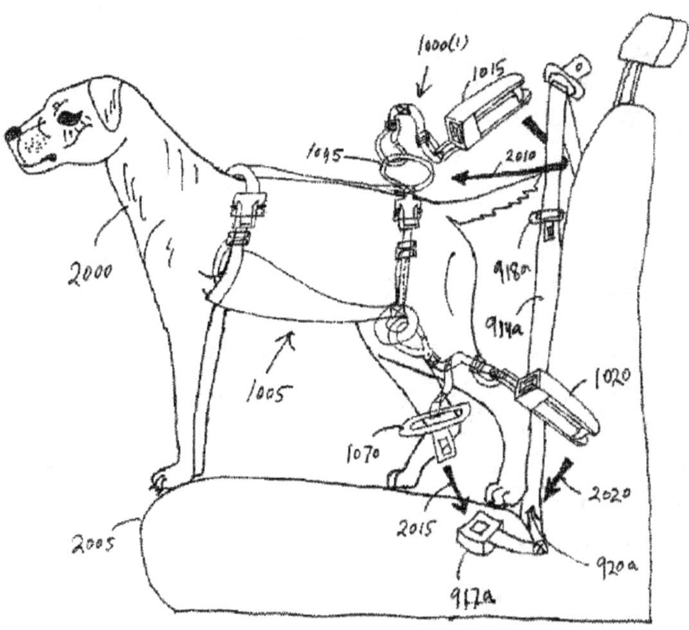

Chapter 6

AUTOMOTIVE BRAKE LIGHTING

Don't DRIVERS driving behind a vehicle need to know how STRONGLY the driver of this vehicle is pressing their brake pedal? Yes, we do. It will HELP us in breaking our vehicle. Agreed.

My this invention (U.S. Patent No. 7,755,474) relates generally to Automotive brake lighting. This invention relates generally to automotive systems, and more particularly, to automotive brake lighting systems.

BACKGROUND OF PROBLEM

A lighting system of a motor vehicle consists of lighting and signaling devices mounted or integrated to the front, sides and rear of the vehicle. The purpose of this system is to provide illumination by which for the driver to operate the vehicle safely after dark, to increase the conspicuity of the vehicle, and to display information about the vehicle's presence, position, size, direction of travel, and driver's intentions regarding direction and speed of travel.

A lighting system of a motor vehicle also comprises rear position lamps (tail lamps) sometimes referred to as a brake light or a stop light. A brake light or a stop light is a red light on the rear of a motor vehicle that signals when the brakes are applied to slow or stop. Besides the conventional taillights sometimes A Center High-Mounted Stop Lamp (usually seen abbreviated as CHMSL) is used as a third stop lamp, or brake light. It is mounted on the rear of a vehicle. It is usually placed above the rear window, or is affixed inside the window and projects through it. In some creative arrangements, the CHMSL is integrated into a spoiler. A CHMSL is usually thought of as a car safety feature.

Nighttime vehicle conspicuity to the rear is provided by rear position lamps disposed in a brake light, stop light, tail light, tail lamp and rear light. These are required to produce only red light, and to be wired such that they are lit whenever the front position lamps are illuminated --including when the headlamps are on. Rear position lamps may be combined with the vehicles brake lamps, or separate from them. In combined-function installations, the lamps produce brighter red light for the brake lamp function, and dimmer red light for the rear position lamp function. Regulations worldwide stipulate minimum intensity ratios between the bright (brake) and dim (tail) modes, so that a vehicle displaying rear position lamps will not be mistakenly interpreted as showing brake lamps, and vice versa. Rear position lamps are permitted, required or forbidden to illuminate in combination with daytime running lamps, depending on the jurisdiction and the DRL implementation.

SOLUTION

Generally, a method and apparatus are provided for controllably illuminating automotive brake lights. In one embodiment, an automotive brake lighting system comprises a brake light including first and second light sources, a sensor to sense a level of an operating condition associated with at least one of a brake assembly or a brake pedal of a motor vehicle for illuminating the brake light, and a controller configured to monitor information provided by the sensor. The controller, in response to the level of the operating condition, selectively powers the at least one of the first or second light sources of the brake light based at least in part on the monitored information to generate a variable visual indication relating to braking of the motor vehicle from the brake light over a range of at least two different visual indications that indicate a variable braking rate of the motor vehicle over a corresponding range of at least two different braking rates.

In one embodiment of the present invention, a method is provided for illuminating a brake light including first and second light sources. The method comprises monitoring information provided by a sensor that senses a level of an operating condition associated with at least one of a brake assembly or a brake pedal of a motor vehicle. The method further comprises, in response to the level of the operating condition, selectively powering the at least one of the first or second light sources of the brake light based at least in part on the monitored information to generate a variable visual indication relating to braking of the motor vehicle from the brake light over a range of at least two different visual indications that indicate a variable braking rate of the motor vehicle over a corresponding range of at least two different braking rates.

Consistent with one illustrative embodiment of the present invention, the automotive brake lighting system 102 any further comprise memory 150 coupled to the controller 115. The memory 150 may be capable of storing power control information 155 associated with a power signal 160. The controller 115 may be capable of retrieving the power control information 155 for selectively powering the first and/or second light sources 110(1,2) of the first brake light 105(1).

In operation, to selectively power the first and/or second light sources 110(1, 2) of the first brake light 105(1), the controller 115 may be configured to monitor information (INFO) 165 provided by the sensor 130. Based at least in part on the monitored INFO 165, in response to a level of the operating condition, the controller 115 may selectively power either one of the first light source 110(1) or the second light source 110(2) or both of them at the same time.

By selectively lighting or switching ON one or both of the first and second light sources 110(1,2) of the first brake light 105(1), the controller 115 may generate a variable visual indication relating to braking of the motor vehicle 100 from the first brake light 105(1) over a range of at least two different visual indications 170(1,2). These two different visual indications 170(1,2) may indicate a variable braking rate of the motor vehicle 100 over a corresponding range of at least two different braking rates 175(1,2). For example, in accordance with one embodiment of the present invention, a braking rate may refer to a change in revolutions per second of a tire 180 of the motor vehicle 100. That is, the braking rate may be defined based on a number of revolutions per second by which the tire 180 of the motor vehicle 100 may be rotating or braking. In other words, the braking rate may correlate to slowing down of the tire 180 as indicated by a desired speed level in miles per hour shown by a speedometer or a desired distance measured in feet that the motor vehicle 100 may traverse as shown by a trip meter on a road or a similar pavement surface and the like before coming to a partial or complete halt.

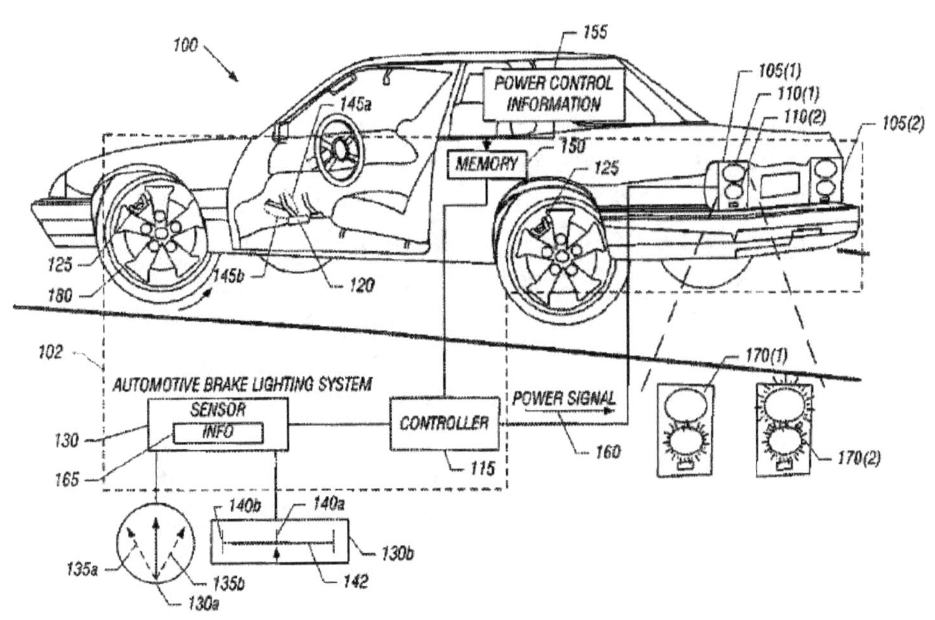

Chapter 7

CUSTOMIZABLE AUDIO EARPIECE

Don't persons with SMALL ear canals need to listen to music while exercising or jogging? Indeed, they do. Agreed.

My this invention (U.S. Patent No. 8,270,654) relates generally to an electronic earpiece device and more particularly, to a customizable audio earpiece.

BACKGROUND OF PROBLEM

Use of audio earpieces that include an ear bud which is configured to be received in the ear of a user of a personal electronic device is well known. While many of such audio earpieces adequately produce sound from playing prerecorded audio content, they come with a standard size ear buds. The ear canal cavity of the ear supports such standard size ear bud during their usage. In particular, a typical ear bud has a fixed stick-shaped support to orient the ear bud within the conchal bowl of the ear.

However, users have varying size of ears with different size of ear canal cavities and conchal bowls. In other words, since no two ears are created equal, the approach of one size fits all is somewhat not ideal. This approach makes the experience of extended listening to a favorite music or audio content sometimes a bit painful and tiring. Likewise, using Bluetooth headsets with cell phones can be cumbersome experience. This is a particular problem where the users either have smaller than normal size ears or have slightly bigger ears with oddly shaped conchal bowls.

SOLUTION

A customizable audio earpiece includes an ear bud body with an ear bud size of which is adjustable along with some control over its orientation. The ear bud body comprises a ball point with teeth that couples to a mating cavity with teethed groves to movably lock or snap with each other on an ear bud stick. The flexibility of one or more petal shaped fingers provides an adequate level of stiffness for the ear bud when resting therein. The teeth on the ends of these petal shaped fingers releasably couple to teeth on a speaker body. In this way, users have varying size of ears with different size of ear canal cavities and conchal bowls can comfortably use the customizable audio earpiece.

In one embodiment of the present invention, a customizable audio earpiece for use with a personal electronic device to fit in a first ear of a user having a first ear canal cavity is provided. The customizable audio earpiece may comprise a first ear bud body including a first speaker and configured to be supported by the first ear. The first ear bud body may include a first ear bud having a face, a major end and a minor end, a first ear bud support coupled to the minor end of the first ear bud and a plurality of first generally flexible fingers coupled to the first ear bud at the minor end and extending over the first speaker towards the major end of the first ear bud for supporting the first ear bud by the first ear canal cavity.

In another embodiment of the present invention, a customizable audio earpiece for use with a personal electronic device to fit in a first ear of a user having a first ear canal cavity is provided. The customizable audio earpiece may comprise a first ear bud body including a first speaker and configured to be supported by the first ear. The first ear bud body may include a first ear bud having a face, a major end and a minor end and a first ear bud support coupled to the minor end of the first ear bud, wherein the first ear bud further comprising a first component at the minor end, and the first ear bud support including a second component to movably receive the first component such that the first ear bud is adjustable relative to the first ear bud support to at least two positions.

In yet other embodiment of the present invention, a customizable audio earpiece for use with a personal electronic device to fit in an ear of a user having an ear canal cavity is provided. The customizable audio earpiece may comprise an ear bud body including a speaker and configured to be supported by the ear. The ear bud body may include an ear bud having a front face, a side face, a major end and a minor end, an ear bud support coupled to the minor end of the ear bud and a plurality of generally flexible fingers coupled to the ear bud and extending over the speaker for supporting the ear bud by the ear canal cavity.

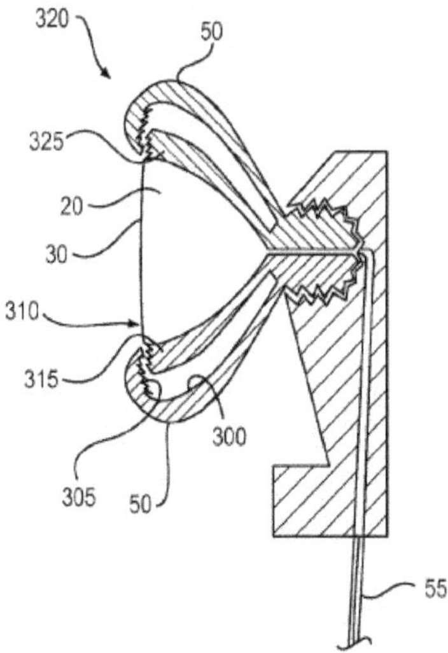

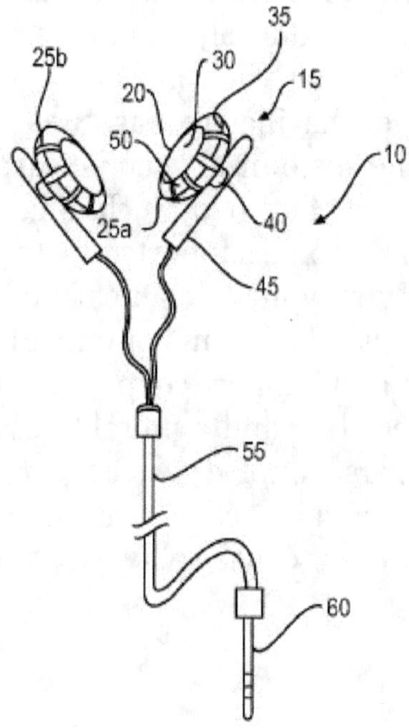
35
25b
30
15
20
50
10
25a
40
45
55
60

Chapter 8

HAND-HELD, PORTABLE ELECTRONIC DEVICE WITH RETAINER PORT FOR RECEIVING RETAINABLE WIRELESS ACCESSORY FOR USE THEREWITH

Don't we all need wireless accessories to STOW AWAY with a mother device for later use or just retaining? Indeed, we do. Agreed.

My this invention (U.S. Patent No. 8,472,658) relates generally to hand-held, portable electronic devices, and more particularly, to providing a retainer in a hand-held, portable electronic device for retaining a wireless accessory.

BACKGROUND OF PROBLEM

Increasingly use to hand-held, portable electronic devices, such as a processor or controller based devices including hand-held, portable computers, hand-held multi-media players, music players, cellular phones, hand-held wired and/or wireless communication and/or computing devices, hand-held pocket computers, and personal digital assistants is becoming widely popular. As a result, use and exchange of electronic entertainment content and information is not only a trend anymore but a norm and a convenient way of experiencing multi-media content such as music and video, conveying information including electronic mail messages between users of wired and/or mobile communication devices. Many commercial enterprises, media service providers or network operators. Internet service providers and businesses use Internet to disseminate electronic multi-media content (textual, graphic, music and video files) over a connected mesh of wired and/or wireless network users. For example, several of these entities usually provide access to multi-media content and services on the Internet via websites and web browsers.

Users of the hand-held, portable electronic devices obtain and store a variety of electronic multi-media content such as music and video files on their hand-held, portable devices. Likewise, users of networked devices wired and/or wireless exchange electronic messages to communicate with other users. To use the hand-held, portable electronic devices, an audio-phone comprising headphone(s) and/or microphone is provided for users. In general, for listening to electronic multi-media content such as music and video files, a pair of stereo headphones is provided with the hand-held, portable electronic devices. For the hand-held, portable networked, wired and/or wireless communication and/or computing devices, a headphone with a microphone is provided. Examples of headphones and/or microphones include a wired or a wireless set of audio devices.

While the wireless headphones and/or microphones comprise a headphone and/or a microphone, the wired headphone(s) and/or microphone comprise a pair of stereo audio cords with a set of micro headphones at one end and an audio male connector pin on the other end, both face a similar set of issues relating to their storage for reuse, i.e., when not being used by users. In particular, the audio cords are a few feet long so that a user can conveniently use the headphones and/or microphones. Since these audio cords are relatively soft and flexible they get entangled when the headphones and/or microphones are not being used. That is, storage of the headphones and/or microphones with a relatively long and flexible audio cord with headphone speaker(s) attached to its end is a messy affair. With regard to the wireless headphones and/or microphones, because of their wireless convenience it is relatively easy to misplace, damage, or loose them when they are not being in use. In this manner, storage for reuse of both the wired and/or wireless headphones and/or microphones of the handheld, portable electronic devices becomes quite a challenge when an audio-phone is not put to use by wireless users or it discharges rather quickly due to a constant use and need frequent recharging.

A wireless device accessory is a wireless communication-enabled accessory that can electronically, i.e., wirelessly communicate with a wireless communication-enabled device over a wireless communication link.

With regard to wireless device accessories such a stylus or a Bluetooth headset, because of their wireless convenience it is relatively easy to misplace, damage, or loose them when they are not being in use. In this manner, storage for reuse of wireless device accessories of the hand-held, portable electronic devices becomes quite a challenge when a wireless device accessory is not put to use by users.

SOLUTION

A hand-held, portable electronic device is provided with a retainer port that defines an accessory retaining structure formed in an exterior periphery of a device housing for removably retaining a retainable wireless accessory in a cavity of the accessory retaining structure. In one embodiment of the present invention, the hand-held, portable electronic device includes a first transceiver adapted to communicate with the retainable wireless accessory over a short-range wireless communication link. The retainable wireless accessory may include a second transceiver adapted to communicate with the hand-held, portable electronic device using the short-range wireless communication link.

In one embodiment of the present invention, a hand-held, portable electronic device for use with a retainable wireless accessory including a conductive member is provided. The hand-held, portable electronic device comprises a device body and a device housing that encloses the device body. The device housing has an exterior periphery. The hand-held, portable electronic device further comprises a retainer port that defines an accessory retaining structure formed in the exterior periphery of the device housing, the accessory retaining structure is configured to receive the conductive member of the retainable wireless accessory for storing the retainable wireless accessory substantially external to the exterior periphery of the device housing of the hand-held, portable electronic device, wherein the hand-held, portable electronic device includes a first transceiver adapted to communicate with the retainable wireless accessory over a wireless communication link, the retainable wireless accessory including a second transceiver adapted to communicate with the hand-held, portable electronic device using the wireless communication link, wherein the accessory retaining structure including: a cavity for retaining the retainable wireless accessory at the hand-held, portable electronic device by inserting the conductive member of the retainable wireless accessory within the cavity, wherein the conductive member of the retainable wireless accessory to enable flow of charge such that the retainable wireless accessory can function by wirelessly communicating with the hand-held, portable electronic device when physically not connected thereto, wherein the retainable wireless accessory is stored for reuse when physically connected to the hand-held, portable electronic device, wherein the accessory retaining structure is configured to accept a male member which extends longitudinally from an end of the retainable wireless accessory.

In another embodiment of the present invention, a method of retaining at a hand-held, portable electronic device a retainable wireless accessory including a conductive member is provided. The method comprises providing a device body of the hand-held, portable electronic device, providing a device housing that encloses the device body, the device housing having an exterior periphery and providing a retainer port that defines an accessory retaining structure formed in the exterior periphery of the device housing, the accessory retaining structure is configured to receive the conductive member of the retainable wireless accessory for storing the retainable wireless accessory substantially external to the exterior periphery of the device housing of the hand-held, portable electronic device, wherein the hand-held, portable electronic device includes a first transceiver adapted to communicate with the retainable wireless accessory over a wireless communication link, the retainable wireless accessory including a second transceiver adapted to communicate with the hand-held, portable electronic device using the wireless communication link, wherein the accessory retaining structure including: a cavity for retaining the retainable wireless accessory at the hand-held, portable electronic device by inserting the conductive member of the retainable wireless accessory within the cavity, wherein the conductive member of the retainable wireless accessory to enable flow of charge such that the retainable wireless accessory can function by wirelessly communicating with the hand-held, portable electronic device when physically not connected thereto, wherein the retainable wireless accessory is stored for reuse when physically connected to the hand-held, portable electronic device, wherein the accessory retaining structure is configured to accept a male member which extends longitudinally from an end of the retainable wireless accessory.

In yet another embodiment of the present invention, a kit is provided that comprises a hand-held, portable electronic device and a retainable wireless accessory including a conductive member. The retainable wireless accessory is configured to be used with the hand-held, portable electronic device. The hand-held, portable electronic device includes a device body and a device housing that encloses the device body. The device housing has an exterior periphery. The hand-held, portable electronic device further includes a retainer port that defines an accessory retaining structure formed in the exterior periphery of the device housing, the accessory retaining structure is configured to receive the conductive member of the retainable wireless accessory for storing the retainable wireless accessory substantially external to the exterior periphery of the device housing of the hand-held, portable electronic device, wherein the hand-held, portable electronic device includes a first transceiver adapted to communicate with the retainable wireless accessory over a wireless communication link, the retainable wireless accessory including a second transceiver adapted to communicate with the hand-held, portable electronic device using the wireless communication link, wherein the accessory retaining structure including: a cavity for retaining the retainable wireless accessory at the hand-held, portable electronic device by inserting the conductive member of the retainable wireless accessory within the cavity, wherein the conductive member of the retainable wireless accessory to enable flow of charge such that the retainable wireless accessory can function by wirelessly communicating with the hand-held, portable electronic device when physically not connected thereto, wherein the retainable wireless accessory is stored for reuse when physically connected to the hand-held, portable electronic device, wherein the accessory retaining structure is configured to accept a male member which extends longitudinally from an end of the retainable wireless accessory.

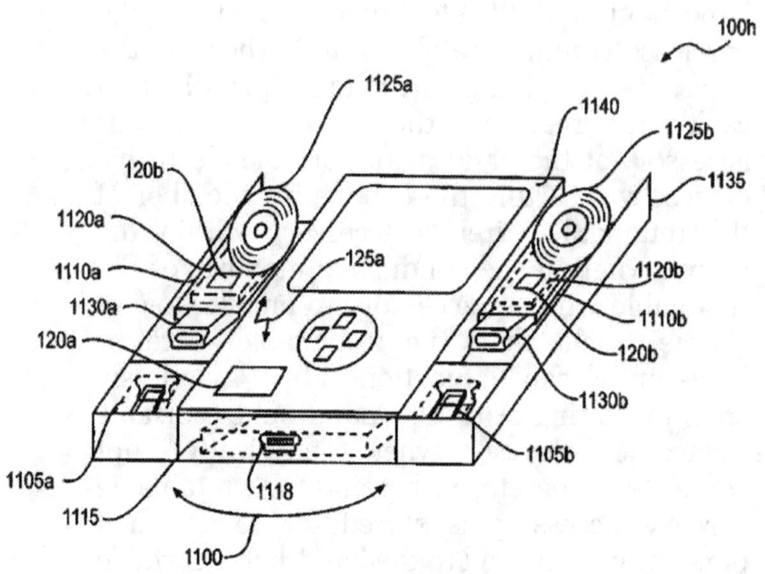
100h
1125a
1140
1125b
120b
1135
1120a
125a
1110a
1120b
1130a
1110b
120a
120b
1130b
1105b
1105a
1118
1115
1100

Chapter 9

PORTABLE DOCUMENT OR NOTEBOOK HOLDER FOR USE WITH PORTABLE COMPUTERS SUCH AS NETBOOK OR LAPTOP

Don't we all elementary school students need to type a hand-written story or project in a computer even when we are not that good in the typing skill? Indeed, we do. Agreed.

My this invention (U.S. Patent Application Publication No. 2013/0221180) relates generally to a document/notebook holder for use with portable computers such as a netbook or a laptop as an accessory to hold either a document or a notebook in front of a user of a portable computer.

BACKGROUND OF PROBLEM

In our this new information age, computers such as netbooks or laptops are present almost everywhere, e.g., in schools, workplaces and homes so typing has become an integral part of daily living for most people including students and workers.

In some of above set forth situations, typing may be done from existing or already written up content. And normally such typing is generally done by people by constantly looking sideways at their self-created content on a paper/document or in a notebook. For example, most school going kids type their hand-written notes including stories and projects etc. on a portable computer, e.g., a netbook or a laptop computer in a similar fashion that involves continuous moving of their necks.

However, such typing being done by repetitively looking sideways on the hand-written notes makes the neck or back of a typist hurt with a pain due to the strain that results from constantly turning the neck to a side for reading what to type on the netbook or the laptop computer.

SOLUTION

A holder such as a portable document/notebook holder is provided for use with a portable computer such as a netbook or a laptop computer to locate a document or a notebook in a front space relative to a user by mounting the holder on to the portable computer and attaching the document or notebook to that holder. Connectors such as clips may be coupled to ends of two arms that are movably coupled to each other on the other ends. For example, a first arm may fold against a second arm in a closed position and the first arm may maintain an open position when the first arm is either partially folded or not folded. For reading hand written notes, instead of repetitive turning of the neck sideways only upward and downward movement of eyeballs is sufficient which avoids straining of neck or back, making the typing much faster and a comfortable task.

In one exemplary embodiment of the present invention, a portable holder is provided for use with a portable computer having a keyboard and a display screen with an edge. The portable holder includes a first arm having a first distal end and a first arm end with a first attachment. The portable holder further includes a second arm having a second distal end and a second arm end with a second attachment. The portable holder further includes a first connector coupled to the first distal end of the first arm, wherein the first connector may be disposed in a first orientation relative to the first and second arms. The first connector may be adapted to removably hold at least one of a document or a notebook. The portable holder further includes a second connector coupled to the second distal end of the second arm, wherein the second connector may be disposed in a second orientation relative to the first orientation of the first connector. The second connector may be adapted to removably mount to the display screen of the portable computer. The first attachment of the first arm may be coupled to the second attachment of the second arm such that the first arm is movably connected to the second arm.

In another exemplary embodiment of the present invention, a portable document/notebook holder is provided for use with a netbook or a laptop computer having a keyboard and a display screen with a top edge. The portable document/notebook holder includes a first arm having a first distal end and a first arm end with a first attachment. The portable document/notebook holder further includes a second arm having a second distal end and a second arm end with a second attachment. The portable document/notebook holder further includes a first clip coupled to the first distal end of the first arm, wherein the first clip may be disposed in a first orientation relative to the first and second arms. The first clip may be adapted to removably hold at least one of a document or a notebook having a given user created or existing content. The portable document/notebook holder further includes a second clip coupled to the second distal end of the second arm, wherein the second clip may be disposed in a second orientation relative to the first orientation of the first clip. The second clip may be adapted to removably mount to the display screen of the portable computer such that, in use, the first orientation of the first clip and the second orientation of the second clip enable the at least one of the document or the notebook to face a given typist present in front of the netbook or the laptop computer when being attached to the first clip and while the second clip is attached to the display screen. The portable document/notebook holder further includes a folding assembly configured and arranged to movably couple the first attachment of the first arm to the second attachment of the second arm such that the first arm is capable of folding against the second arm in a closed position and the first arm is capable of maintaining an open position relative to the second arm when the first aim is at least one of partially or not folded against the second arm. The first arm may have a first predefined length and the second arm may have a second predefined length. The first and second predefined lengths may be selected to position the at least one of the document or the notebook at a desired location relative to the top edge of the display screen.

In another exemplary embodiment of the present invention, a method of holding at least one of a document or a notebook on a portable computer having a keyboard and a display screen with a top edge is provided. The method includes providing a holder including first and second arms that are movably coupled to each other and each of the first and second arms having a clip attached thereto, wherein one of the clips is adapted and oriented to removably hold the at least one of the document or the notebook and other one of the clips is adapted and oriented to removably attach to the top edge of the display screen. The method further includes enabling the first arm to fold against the second arm in a closed position. The method further includes enabling the first arm to maintain an open position relative to the second arm when the first arm is at least one of partially or not folded against the second arm.

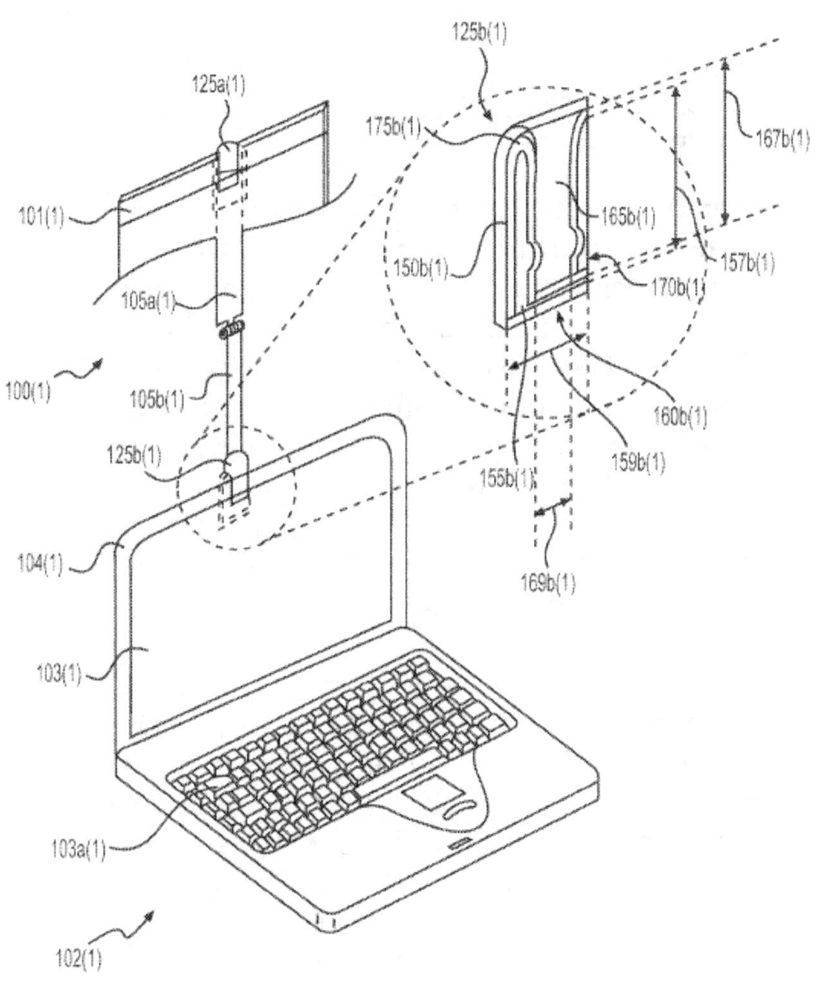
125b(1)
125a(1)
175b(1)
167b(1)
101(1)
165b(1)
150b(1)
157b(1)
170b(1)
105a(1)
100(1)
105b(1)
160b(1)
125b(1)
159b(1)
155b(1)
104(1)
169b(1)
103(1)
103a(1)
102(1)

Chapter 10

VEHICLE ATTACHABLE CHILD BOOSTER TYPE CAR SEAT WITH LAP BELT

Parents in America spend many hundred million dollars each year on car seats including booster type car seats that may not add much lifesaving value. That is, the current booster seats may not be the safety miracle device that most parents have been taught to believe. But children should always ride with some kind of restraint--and a booster seat may be the only legal option. If true, Agreed.

My this invention (U.S. Patent Application Publication No. 2010/0033000) relates generally to child booster type car seats and more particularly, to child booster type car seats with a lap belt that are capable of attaching to a vehicle seat belt system while fully retaining a seat belt functionality.

BACKGROUND OF PROBLEM

National Highway Traffic Safety Administration (NHTSA) recommends that rear-facing seats be used in the back seat from birth to at least 1 year old and at least 20 pounds. Forward-facing toddler seats to be used in the back seat from age 1 and 20 pounds to about age 4 and 40 pounds. Booster seats to be used in back seat from about age 4 to until they are at least 8 years old, unless they are 4'9" tall. Safety seat belts to be used at age 8 and older or taller than 4'9". All children age 12 and under should ride in the back seat of a vehicle.

Booster seats are used because safety seat belts in vehicles are not designed for children. Beginning at about 4 years old age, many children outgrow toddler seats but still they are too small for adult-sized safety seat belts. A booster seat raises the child up so that a safety seat belt properly fits--and can better protect the child. A three-point safety belt of most modern vehicles includes a shoulder and a lap belt. The shoulder belt should cross the child's chest and rest snugly on the shoulder, and the lap belt should rest low across the pelvis or hip area. Most 4 to 8 years old children need booster seats. National Highway Traffic Safety Administration (NHTSA) recommends that children who have outgrown child safety seats should be properly restrained in booster seats until they are at least 8 years old, unless they are 4'9" tall.

For children younger than roughly 24 months, safety seat belts plainly won't do. For them, a car seat represents the best practical way to ride securely. But for older children would safety seat belts afford them the same protection as car seats?

Fatality Analysis Reporting System (FARS) compiles police reports on all fatal crashes in the U.S. since 1975. There data includes different variables in a crash, including whether the occupants were restrained and how. The FARS data reveals that among children 2 and older, the death rate is no lower for those traveling in any kind of car seat than for those wearing safety seat belts. According to the FARS data, there is no evidence that car seats do a better job than seat belts in saving the lives of children older than 2. In certain kinds of crashes--rear-enders, for instance--car seats actually perform worse. The answer to why child auto fatalities have been falling seems to be that more and more children are restrained in some way. Many of them happen to be restrained in car seats, since that is what the government mandates, but if the government instead mandated proper safety seat belt use for children, they would likely do just as well, without much expense, regulation and anxiety associated with car seats.

NHTSA recommends that all older children (usually starting at about age 4) ride in booster seats, which boost a child to a height where the adult lap-and-shoulder belts fit properly. In 2001, the Insurance Institute for Highway Safety sent NHTSA a memo warning that its booster-seat recommendations were "getting ahead of science and regulations" and that certain booster seats "did not improve belt fit, and some actually worsened the fit." If booster seats are shown in the FARS data to be no more effective than safety seat belts, might it be because so many of them are improperly installed?

Most child-safety experts advocate that neck or abdominal injuries are worse with use of safety seat belts alone. That is, one 4-year-old in a lap-and-shoulder belt may find the shoulder belt so irritating that he puts it behind his back and another 4-year-old may be in a poorly installed booster seat.

SOLUTION

A vehicle attachable child booster type car seat is provided with a lap belt for in situ restraining the car seat at a vehicle seat by using a vehicle safety seat belt having a first vehicle seat belt buckle and a first vehicle seat belt prong. In one embodiment of the present invention, a booster seat is configured to have a second vehicle seat belt buckle and a second vehicle seat belt prong to mate with the first vehicle seat belt prong and the first vehicle seat belt buckle, respectively, when positioned on the vehicle seat of a motor vehicle. Therefore, instead of the booster seat being merely an unattached cushion boosting a child occupant's position, may rather be anchored at the vehicle seat while the vehicle safety seat belt's functionality still retained along with an additional restraining capability of the lap belt that can secure a torso portion of the child occupant seated in the booster seat. Such anchoring of a body of the booster seat to a vehicle seat belt system near a seating base of the vehicle seat may avoid dislodging of a seat pan of the booster seat from the top of the vehicle seat in response to an impact force or sudden breaking of the motor vehicle.

In one embodiment of the present invention, a booster seat is provided for use in a vehicle with a vehicle safety seat belt having a first vehicle seat belt buckle and a first vehicle seat belt prong. The booster seat comprises a seat body having first and second distal ends for seating a child occupant thereon. The booster seat further comprises a second vehicle seat belt buckle coupled to the seat body for mating with the first vehicle seat belt prong to restrain the booster seat at a vehicle seat. The booster seat further comprises a second vehicle seat belt prong coupled to the seat body for mating with the first vehicle seat belt buckle coupled to the vehicle seat such that the booster seat attachable in situ to the vehicle seat at the first distal end.

In another embodiment of the present invention, a child car seat comprises a seating portion to boost a child occupant up so a shoulder belt and a lap belt of a vehicle safety seat belt fit the child occupant when restrained therewith. The vehicle safety seat belt having a first vehicle seat belt buckle and a first vehicle seat belt prong. The child car seat further comprises a back-rest portion coupled to the seating portion. A second vehicle seat belt buckle and a second vehicle seat belt prong coupled to the child car seat for coupling to the vehicle safety seat belt.

In yet other embodiment of the present invention, a method is provided for enabling restraining of a booster seat in a vehicle with a vehicle safety seat belt having a first vehicle seat belt buckle and a first vehicle seat belt prong. The method comprises providing the booster seat comprising a body including a seat back and a seat pan for seating a child occupant thereon. The method further comprises enabling coupling a second vehicle seat belt buckle to the body for mating with the first vehicle seat belt prong to restrain the booster seat at a vehicle seat. The method further comprising enabling coupling a second vehicle seat belt prong to the body for mating with the first vehicle seat belt buckle coupled to the vehicle seat such that the booster seat in situ attachable to the vehicle seat.

Duality Innovation

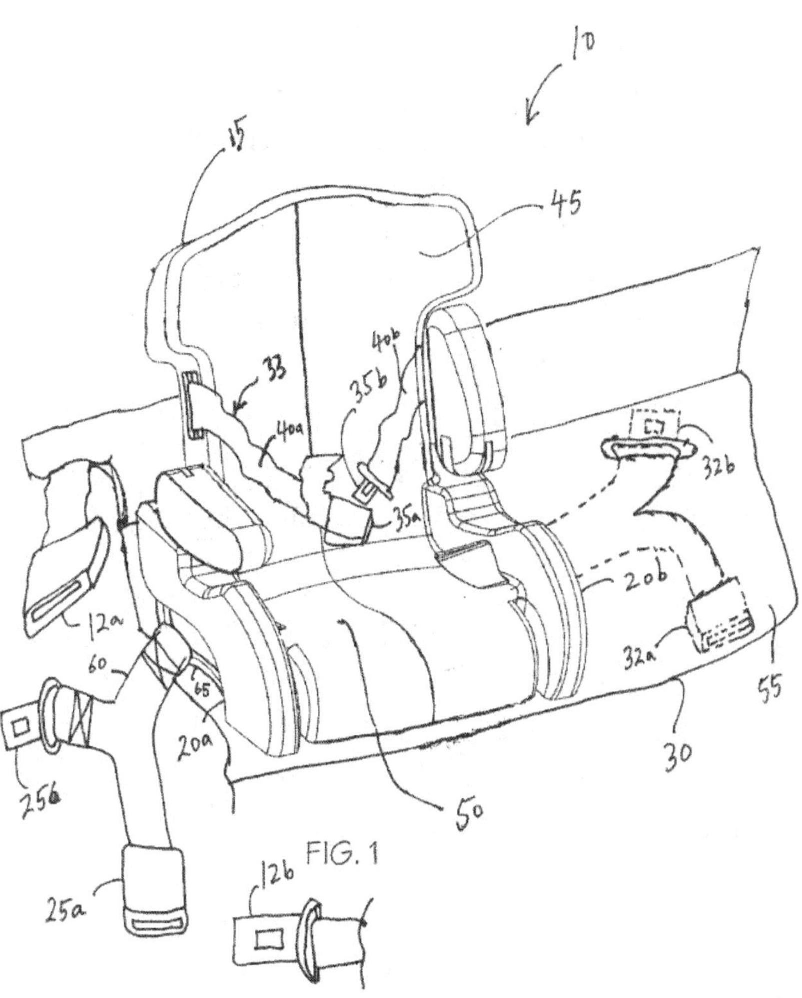

FIG. 1

Chapter 11

LOCKING-OUT A DRIVER HANDHELD MOBILE DEVICE DURING DRIVING OF A VEHICLE FOR TEXTING AND BROWSING

Driving and texting or browsing don't mix well. People die in thousands every year. Agreed.

My this invention (U.S. Patent No. 9,661,127) relates generally to safe operation of handheld mobile devices, and more particularly, to providing a lock-out mechanism to prevent operation of one or more functions of handheld mobile devices by drivers when operating vehicles.

BACKGROUND OF PROBLEM

When you are driving, how often do you see other drivers checking their phones while behind the wheel? And, be honest, how often do you do it yourself? The problem has gotten so big that highways across the country now regularly warn drivers "Don't text and drive." And 46 states and the District of Columbia have laws banning texting and driving. But the issue isn't just talking and texting anymore. Drivers are on Snapchat, Facebook, Instagram,

Twitter, Google Maps, Spotify, YouTube and now “Pokémon Go,” the video game that has captured the world's attention and has become the latest concern for distracted-driving advocates.

Judging by the results of a recent survey, we have a long way to go in getting that message out. Brutally Honest: How to keep your teens from texting and driving. Nearly 70% of teens say they use apps while driving, according to a just-released survey of 2,500 high school-age children across the country. When the teens were asked to rank the behaviors, they consider the most distracting or dangerous for a teen driver, 29% said driving under the influence of alcohol and 25% said writing or sending a text message. Only 6% said actively looking at or posting to social media is the most distracting or dangerous behavior behind the wheel for a teen driver, according to the survey by Liberty Mutual Insurance and Students Against Destructive Decisions (SADD). In another survey (PDF), this one sponsored by the National Safety Council and focusing on 2,400 drivers of all ages, 74% said they would use Facebook while driving, and 37% said they would use Twitter while behind the wheel, with YouTube (35%) and Instagram (33%) close behind.

Texting while driving has become a major concern of parents, law enforcement, and the general public. An April 2006 study found that 80 percent of auto accidents are caused by distractions such as applying makeup, eating, and text messaging on handheld mobile devices (texting). According to the Liberty Mutual Research Institute for Safety and Students Against Destruction Decisions, teens report that texting is their number one distraction while driving. Teens understand that texting while driving is dangerous, but this is often not enough motivation to end the practice.

New laws are being written to make texting illegal while driving. However, law enforcement officials report that their ability to catch offenders is limited because the texting device can be used out of sight (e.g., on the driver's lap), thus making texting while driving even more dangerous. Texting while driving has become so widespread it is doubtful that law enforcement will have any significant effect on stopping the practice.

SOLUTION

Lock-out mechanisms for driver handheld mobile devices are provided to prevent operation of one or more functions of handheld mobile devices by drivers when operating vehicles based on device location data and device motion data within a vehicle by wirelessly communicating with at least one wireless station access point located inside of the vehicle. The lock-out mechanisms disable the ability of a handheld mobile device to perform certain functions, such as texting or browsing, while one is driving. In one embodiment, a handheld mobile device may provide a lock-out mechanism without requiring any modifications or additions to a vehicle by using a motion analyzer, a mobile device position locator and a lock-out mechanism. In other embodiments, the handheld mobile device may provide a lock-out mechanism with modifications or additions to the vehicle, including the use of wireless signals transmitted by the vehicle or by a wireless station access point disposed within the vehicle.

Briefly described, aspects of the present invention relate to lock-out mechanisms for driver handheld mobile devices. In particular, a lock-out mechanism disables the ability of a handheld mobile device to perform certain functions, such as texting and/or browsing, while one is driving a vehicle. A short-range wireless such as Bluetooth and/or Wi-Fi-based wireless local area network (WLAN) system comprises a motion analyzer to detect whether a vehicle or a handheld mobile device is in motion beyond a predetermined threshold level. The short-range wireless Bluetooth and/or Wi-Fi-based WLAN system further comprises a mobile device position locator configured to determine whether the handheld mobile device is located within a safe operating area of the vehicle based on a wireless signal metric associated with the handheld mobile device using a Bluetooth and/or WLAN station operating as an access point which may be a dongle. The short-range wireless Bluetooth and/or Wi-Fi-based WLAN system further comprises a lock-out mechanism configured to automatically and selectively disable one or more functions of the handheld mobile device based on outputs from the motion analyzer and the mobile device position locator. One of ordinary skill in the art appreciates that such a handheld mobile device position locator and temporarily disabler vehicle system can be configured to be installed in different environments where texting and browsing user communications on a handheld mobile device are used in a vehicular setup when operating vehicles, for example while seating in a driver seat and driving a vehicle using interactive device functions of a handheld mobile device manually by typing in a user interface of the device or scrolling displayed items on a device screen or viewing a website, video or photos. Drivers often use a handheld mobile device for texting, Facebook reading while driving, or they use Twitter while behind the wheel, or watch YouTube and browse Instagram when operating vehicles, risking their life and that of other occupants. The lock-out mechanism disables the ability of a handheld mobile device to perform these functions while one is driving a vehicle and sitting in an unsafe operating area and seamlessly enables them when not driving, meaning not seating in the driver seat but seating in the front passenger seat or rear passenger seats which is a safe operating area.

In accordance with one illustrative embodiment of the present invention, a handheld mobile device is provided. The handheld mobile device comprises a motion analyzer configured to detect whether the handheld mobile device is in motion beyond a predetermined threshold level. The handheld mobile device further comprises a mobile device position locator configured to wirelessly determine whether the handheld mobile device is located within a safe operating area of a vehicle based on location data of the handheld mobile device determined from a wireless signal associated with the handheld mobile device. The handheld mobile device further comprises a lock-out mechanism configured to automatically and selectively disable one or more functions of the handheld mobile device based on outputs from the motion analyzer and the mobile device position locator.

Consistent with another embodiment, a method of locking-out a driver handheld mobile device during driving of a vehicle for texting and browsing is provided. The method comprises detecting using a handheld mobile device whether the handheld mobile device is in motion beyond a predetermined threshold level, determining using the handheld mobile device whether the handheld mobile device is located within a safe operating area of a vehicle based on location data of the handheld mobile device determined from a wireless signal associated with the handheld mobile device and automatically and selectively disabling using the handheld mobile device one or more functions of the handheld mobile device when both the handheld mobile device is detected to be in motion beyond the predetermined threshold level and the handheld mobile device is determined not to be located within the safe operating area of the vehicle.

According to yet another embodiment of the present invention, a wireless in-vehicle mobile device positioning and location system is provided for locking-out a driver handheld mobile device during driving of a vehicle for texting and browsing. The system comprises a motion analyzer configured to detect whether the handheld mobile device is in motion beyond a predetermined threshold level. The system further comprises a mobile device position locator configured to wirelessly determine whether the handheld mobile device is located within a safe operating area or an unsafe operating area of a vehicle based on location data of the handheld mobile device determined from a wireless signal associated with the handheld mobile device. The system further comprises a lock-out mechanism configured to automatically and selectively disable one or more functions of the handheld mobile device if determined to be in the unsafe operating area or enable the one or more functions of the handheld mobile device if determined to be in the safe operating area based on outputs from the motion analyzer and the mobile device position locator.

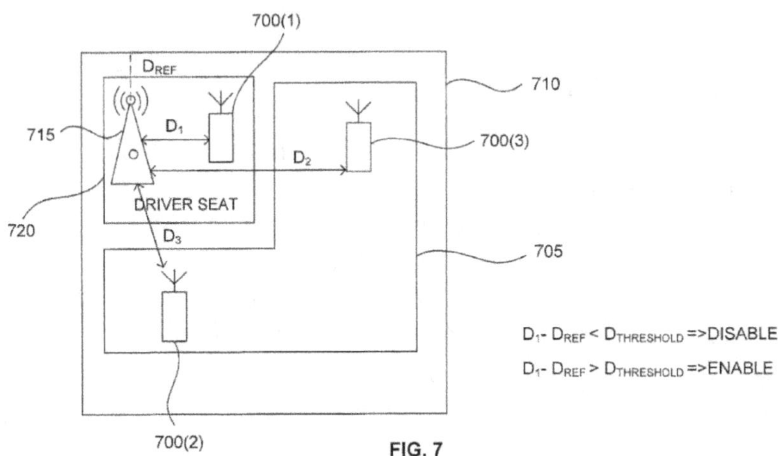

FIG. 7

Chapter 12

ALARMING PARENTS FOR STOPPING CHILDREN BEING LEFT IN A VEHICLE BY MISTAKE

30 plus children dying in hot cars in U.S. every year is not acceptable. Agreed.

My this invention (U.S. Patent No. 9,741,224) relates generally to reminding drivers who might forget that a child is in the vehicle, and more particularly, to providing an alarm to prevent children being left in a vehicle by mistake by a child caregiver for stopping hot vehicle deaths of children due to heat strokes.

BACKGROUND OF PROBLEM

On average, 37 children die from heat-related deaths after being trapped inside hot vehicles. Even the best of parents or caregivers can unknowingly leave a sleeping baby in a car; and the end result can be injury or even death. Often, a parent has forgotten to drop a child off at daycare.

An infant car seat with technology designed to remind drivers of their backseat passengers, and stop children from dying in hot cars is now available. A sensor on the infant seat harness triggers a series of tones if a child is still buckled in when the ignition is switched off. The feature is meant to remind drivers who might forget that a child is in the vehicle. It has a wireless receiver that plugs into a car's on-board diagnostic port and syncs with the chest clip that goes around the baby. It does not require the use of Bluetooth, cellular or other devices. It's the first and only crash-tested car seat that has this type of technology embedded.

Right now (on the market) it's more attachments or accessories or mobile apps, but there's not one that's an actual car seat that has this technology. Hundreds of products invented by well-meaning people to prevent children from dying in a hot car, and the new infant car seat is the most promising development so far. However, the parents have to buy this specific car seat only and cannot use their own existing car seats.

SOLUTION

Reminder mechanisms for driver handheld mobile devices are provided to provide an alarm to prevent children being left in a vehicle by mistake by a child caregiver for stopping hot vehicle deaths of children due to heat strokes.

A wireless in-vehicle reminder device system comprises a wireless in-vehicle reminder device having a housing including a plug to snugly insert into a socket or a port of a vehicle to connect to a power source of the vehicle. The wireless in-vehicle reminder device includes a first wireless transceiver configured to communicate on a first wireless link with a first wireless module of a handheld mobile device and a reminder mechanism configured to provide a reminder in a form of an alarm signal to the handheld mobile device over the first wireless link in response to a warning event detected in the vehicle. The wireless in-vehicle reminder device system further comprises a reminder application (APP) installed on the handheld mobile device and associated with the reminder mechanism to provide at least one of an audio warning and a visual warning from the handheld mobile device in response to the alarm signal.

Briefly described, aspects of the present invention relate to reminder mechanisms which provide an alarm to a driver of a vehicle for not unknowingly leaving a sleeping baby in a vehicle where the end result can be an injury or even death of the child in a hot vehicle due to a heat stroke. In particular, a reminder mechanism is built in a pluggable cellular device having a subscriber identity module or subscriber identification module (SIM) card, a motion sensor, a voice sensor and an on-board battery for providing an alarm signal to an associated reminder application (APP) installed on a mobile device such as a cell phone of the caregiver or the driver of the vehicle. A SIM is an integrated circuit chip that is intended to securely store the international mobile subscriber identity (IMSI) number and its related key, which are used to identify and authenticate subscribers on mobile telephony devices (such as mobile phones and computers). The pluggable cellular device may be a Bluetooth-enabled device with the on-board battery. It gets paired up with a caregiver or driver cell phone via a Bluetooth protocol and may be programmed to send a Short Message Service (SMS) to the reminder APP associated with the pluggable cellular device and installed on the caregiver or driver cell phone. The pluggable cellular device may include a plugging interface to plug into a vehicle standard power outlet/cigarette lighter port. Based on outputs of the voice sensor and the motion sensor which may indicate either motion or audio signal inside the vehicle and detecting that the vehicle ignition is turned off, the pluggable cellular device may transmit an alarm signal and/or a Short Message Service (SMS) to the reminder APP installed on the caregiver or driver cell phone over a cellular link so that the caregiver or driver cell phone emits a loud tone. Short Message Service (SMS) is a text messaging service component of phone, Web, or mobile communication systems.

In accordance with one illustrative embodiment of the present invention, a wireless in-vehicle reminder device comprises a housing including a plug to snugly insert into a socket or a port of a vehicle to connect to a power source of the vehicle. The wireless in-vehicle reminder device further comprises a first wireless transceiver configured to communicate on a first wireless link with a first wireless module of a handheld mobile device. The wireless in-vehicle reminder device further comprises a reminder mechanism configured to provide a reminder in a form of an alarm signal as at least one of an audio warning and a visual warning to the handheld mobile device over the first wireless link in response to a warning event detected in the vehicle.

Consistent with another embodiment, a method of reminding a driver who might forget that a child is in a vehicle. The method comprises providing a housing including a plug to snugly insert into a socket or a port of the vehicle to connect to a power source of the vehicle, providing a first wireless transceiver configured to communicate on a first wireless link with a first wireless module of a handheld mobile device and providing a reminder in a form of an alarm signal as at least one of an audio warning and a visual warning to the handheld mobile device over the first wireless link in response to a warning event detected in the vehicle.

According to yet another embodiment of the present invention, a wireless in-vehicle reminder device system is provided. The system comprises a wireless in-vehicle reminder device having a housing including a plug to snugly insert into a socket or a port of a vehicle to connect to a power source of the vehicle. The wireless in-vehicle reminder device includes a first wireless transceiver configured to communicate on a first wireless link with a first wireless module of a handheld mobile device and a reminder mechanism configured to provide a reminder in a form of an alarm signal to the handheld mobile device over the first wireless link in response to a warning event detected in the vehicle. The system further comprises a reminder application (APP) installed on the handheld mobile device and associated with the reminder mechanism to provide at least one of an audio warning and a visual warning from the handheld mobile device in response to the alarm signal.

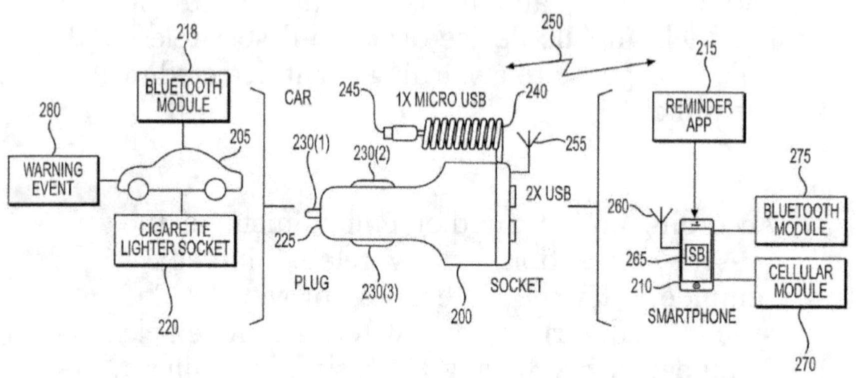
218
BLUETOOTH MODULE
280
WARNING EVENT
205
CIGARETTE LIGHTER SOCKET
220
CAR
245
1X MICRO USB
240
250
215
REMINDER APP
230(1)
230(2)
255
2X USB
275
BLUETOOTH MODULE
260
225
PLUG
230(3)
200
SOCKET
265
SB
210
SMARTPHONE
CELLULAR MODULE
270

Chapter 13

TWO-WHEEL DRIVE BICYCLE WITH A DUAL CHAIN-AND-SPROCKET DRIVE MECHANISM DRIVEN BY A TWO-SIDED PADDLE-SPROCKET SYSTEM

The basic shape and configuration of a typical upright or "safety bicycle", has changed little since the first chain-driven model was developed around 1885. It's about time for a change to make a better bicycle. Agreed.

My this invention (U.S. Patent No. 9,650,106) relates generally to two-wheel drive bicycles, and more particularly, a dual chain-and-sprocket drive mechanism driven by a two-sided paddle-sprocket system.

BACKGROUND OF PROBLEM

A conventional bicycle includes a frame having a normally non-driven front wheel and a rear driven wheel rotatably mounted thereon. The rear wheel conventionally has a driven element thereon which may comprise a rear chain driven sprocket cluster having several different size gears and a rear derailleur mechanism. A pedal driven crankset, that includes a pedal sprocket or sprockets, is mounted on the frame and serves as a prime mover for the drive wheel. A drive chain is trained around one of the pedal sprockets and one of the sprockets of the rear sprocket cluster. The rear derailleur mechanism is manually operable to shift the drive chain to train it about any desired one of the sprockets of the rear sprocket cluster. If the crank set includes two or three pedal sprockets, a front derailleur mechanism is also mounted to shift the chain to train it about any desired one of the pedal gears. Such shifting of the drive chain allows the operator to select an optimum sprocket ratio for driving the bicycle over the terrain on which it is being ridden. These sprocket ratios are commonly referred to as “speeds”. A rear wheel drive bicycle may have rear wheel and pedal sprocket clusters that provide a large number of speeds with 10, 18 and 21 speed bicycles being in widespread use. The frame also includes a seat for supporting the bicycle rider and a front steerably mounted fork on which the front wheel is rotatably mounted. Handle bars are connected to the front steerable fork for steering the bicycle in known manner. The rear and front derailleurs each have controls mounted on the frame in a position to be conveniently reached by the operator to shift speeds while pedaling the bicycle.

Bicycles are increasingly being ridden off of paved roads and on rough terrain, which is steep, unpaved, frequently wet or muddy and covered by vegetation in places. Off road trails are particularly prone to have treacherously slippery sections caused by mud or uncut vegetation such as grass and weeds. Because of the rough and slippery terrain bikes are driven over, increased traction is desired. To achieve such traction, it is known to provide a multi-speed bicycle with a drive mechanism that allows for simultaneous and constant driving of both the front and rear wheels. This type of drive mechanism will be referred to as a constant dual wheel drive.

Bicycles have traditionally operated as a single-wheel-drive vehicle. As is evidenced by their popularity, single-wheel-drive bicycles are suitable in most cases. Because they only employ single-wheel drive, however, the use of bicycles are, for the most part, somewhat limited to prepared surfaces such as paved streets, sidewalks, and groomed paths. Although just about everyone who has ever ridden a bicycle has ridden on gravel or unprepared surfaces, all riders know that it is more difficult to ride on these types of surfaces due to the fact that drive is being generated by only the rear wheel.

In any event, single-wheel-drive bicycles are the norm because of difficulties involved in transferring to the front wheel the drive generated by the rider. The difficulty in generating drive via the front wheel of a bicycle results from the need to allow the fork (on which the front wheel is mounted) freedom to turn substantially in either direction from the center position in order to permit the cyclist to steer the bicycle. Because the fork must be free to turn, it is not possible to directly connect the pedals to the front wheel.

As off-road biking has gained broader appeal, the demands that riders place on their bicycles have increased dramatically. Downhill, snow, and endurance races demand the increased traction and mobility of a two-wheel drive bicycle system. Conventional bicycles are powered through a chain linking the pedal crankshaft to the rear wheel. Bicyclists are now facing many obstacles where having only rear wheel drive can lead to bicycle damage or personal injury. For professional riders, precious race time is lost avoiding obstacles such as logs, rocks, loose sand, mud, or ice. Traction and climbing ability are severely limited in extreme mountain conditions by only having the rear wheel provide power. In fact, biking professionals teach that only through learning to keep your weight on the rear wheel will beginners ever hope to improve their off-road skills. Accordingly, there is a need in the industry for a two-wheeled drive bicycle which efficiently transfers power from the pedals to the front wheels, provides the rider with increased ability to safely negotiate rough terrain, and which does not detract from the aesthetic qualities and appearances of the bicycle structure itself.

The concept of a two-wheel drive bicycle is not a new one. Several two-wheel drive bicycle systems are patented at present using various combinations of chains, flexible cable shafts, and rocker arms mounted on the handlebars to transfer power either directly from the pedals to the front wheel or from the rear wheel to the front wheel. While two-wheel-drive bicycles have been proposed, there remains a need for an easily-operable and readily steerable two-wheel-drive bicycle, so that riders will have greater opportunity for enjoyable cycling on off-road trails and other unprepared surfaces.

SOLUTION

Briefly described, aspects of the present invention relate to a two-wheel-drive bicycle that includes a dual chain-and-sprocket drive mechanism driven by a two-opposite-sided paddle-sprocket system. A two-wheel-drive bicycle has separate chain-and-sprocket drive mechanisms for the front and rear wheels configured to be both chain-and-sprocket drive mechanisms driven by a common paddle system which is standard on typical bicycles. A first chain is mounted on a first sprocket and a rear gear coupled to a rear wheel of the bicycle. A second chain is mounted on a second sprocket and a front gear coupled to a front wheel of the bicycle. Both the first and the second sprockets are coupled to a standard paddle system of the bicycle to be driven by legs and feet of a biker. The standard paddle system has a common shaft on which the first and the second sprockets are mounted. The first and the second sprockets may be mounted on left, right or one left one right configuration with respect to the bicycle sides. The chain-and-sprocket drive mechanisms may be multi-sprocket systems. The bicycle may include a multi-gear system to be driven at different gear ratios for generating different torques. The multi-sprocket systems enable setting different gear ratios for the front and rear wheels, selectively providing a different drive force/torque to both wheels. By providing a torque power separately to both rear and front wheels from the same paddles using the dual chain-and-sprocket drive mechanism, the two-wheel drive bicycle will run more stable and have a better traction control over a road surface just like the 4×4 vehicles.

In accordance with one illustrative embodiment of the present invention, a two-wheel-drive bicycle is provided. The two-wheel-drive bicycle comprises a frame having front and rear portions and a center portion with upper and lower ends. The frame has first and second opposing sides and the frame includes a head tube having a first front sprocket mounted thereon. The two-wheel-drive bicycle further comprises a fork assembly attached to the front portion of the frame and pivotable about a steering axis. The fork assembly has a lower end and an upper end. The two-wheel-drive bicycle further comprises a front wheel rotatably mounted on the fork assembly. The fork assembly straddling and rotatably supporting the front wheel at the lower end. The fork assembly has a stem at the upper end for pivotally mounting the fork assembly in the head tube of the frame to enable steering of the front wheel about the stem. The two-wheel-drive bicycle further comprises a rear wheel rotatably mounted on the rear portion of the frame and having a rear sprocket. The two-wheel-drive bicycle further comprises a paddle assembly having first and second foot paddles. The paddle assembly is mounted at the lower end of the center portion in a shared driving relationship with the front wheel and the rear wheel. The two-wheel-drive bicycle further comprises a chain-driven rear-drive mechanism configured for transmitting rotational power to the rear wheel by alternate pumping of a bicycler's legs and feet which drives the paddle assembly. The chain-driven rear-drive mechanism includes a rear main-drive sprocket at the lower end of the center portion and disposed on the first side of the frame. The chain-driven rear-drive mechanism is coupled to the paddle assembly. The two-wheel-drive bicycle further comprises a chain-driven front-drive mechanism configured for transmitting rotational power to the front wheel by alternate pumping of the bicycler's legs and feet which drives the paddle assembly. The chain-driven front-drive mechanism includes a front main-drive sprocket at the lower end of the center portion and disposed on the second side of the frame being opposite of the first side of the frame. The chain-driven front-drive mechanism is coupled to the paddle assembly.

Consistent with another embodiment, a multi-wheel-drive bicycle is provided. The multi-wheel-drive bicycle comprises a frame having first and second opposing sides and having front, rear and lower ends. A front wheel is rotatably mounted on the front end of the frame and a rear wheel is rotatably mounted on the rear end of the frame. The multi-wheel-drive bicycle further comprises a chain-driven rear-drive mechanism configured for transmitting rotational power to the rear wheel. The chain-driven rear-drive mechanism includes a rear main-drive sprocket disposed on the first side of the frame and located at the lower end of the frame. The multi-wheel-drive bicycle further comprises a chain-driven front-drive mechanism configured for transmitting rotational power to the front wheel. The chain-driven front-drive mechanism includes a front main-drive sprocket disposed on the second side of the frame being opposite of the first side of the frame and located at the lower end of the frame.

According to yet another embodiment of the present invention, a method of transmitting rotational power to a front wheel and a rear wheel of a two-wheel-drive bicycle is provided. The method comprises providing a frame having first and second opposing sides and having front, rear and lower ends, rotatably mounting the front wheel on the front end of the frame, rotatably mounting the rear wheel on the rear end of the frame, providing a chain-driven rear-drive mechanism on the first side of the frame and at the lower end of the frame, providing a chain-driven front-drive mechanism on the second side of the frame being opposite of the first side of the frame and at the lower end of the frame, transmitting rotational power from a paddle assembly to the rear wheel with the chain-driven rear-drive mechanism including a rear main-drive sprocket disposed on the first side of the frame and, located at the lower end of the frame, and transmitting rotational power from the paddle assembly to the front wheel with the chain-driven front-drive mechanism including a front main-drive sprocket disposed on the second side of the frame being opposite of the first side of the frame and located at the lower end of the frame.

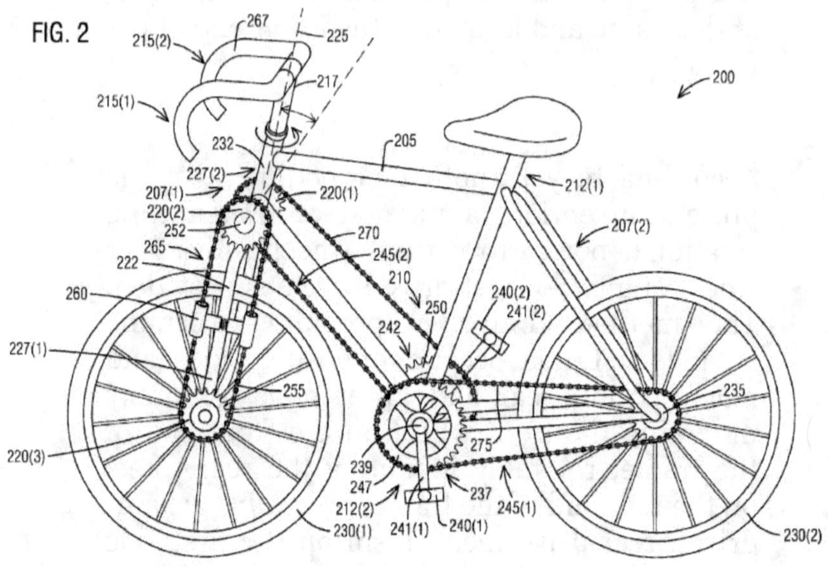
FIG. 2
267
225
215(2)
217
215(1)
200
205
232
227(2)
212(1)
207(1)
220(1)
220(2)
252
270
207(2)
265
245(2)
222
210
240(2)
260
250
241(2)
242
227(1)
255
235
275
220(3)
239
247
237
212(2)
245(1)
230(1)
241(1)
240(1)
230(2)

Chapter 14

WEARABLE SMART WRISTSTRAP OR WATCHBAND WITH INTEGRATED SMARTWATCH FUNCTIONALITY

Arrival of smartwatch should not mean that we will have to abandon our old beloved analog or mechanical wrist watch. Agreed.

My this invention (U.S. Patent Application No. 15/265435) relates generally to a wriststrap or a watchband of a wrist watch or a timepiece, and more particularly, a smart wearable wriststrap or watchband that transforms or converts or turns any regular or classic mechanical or digital timepiece or wrist watch into a hybrid smartwatch.

BACKGROUND OF PROBLEM

Wearable computing has become a prevalent step forward in the progress of technology. Consumers are searching for greater and greater opportunities to integrate technology with everyday wearable items such as glasses, necklaces, and bracelets. Many products on the market today connect to a user's mobile device and allow for the pushing of notifications, answering emails and text messages, as well as the basic functions of keeping time and screening calls.

Mobile, wearables and wellbeing are converging. The data derived from the convergence is starting to enable a new "intimate data economy" analogous to the "digital economy" that sprung up in the mid-1990s. It will enable tackling of 21st century disease. In addition to helping prevent us getting sick over our lifetime it will enable health to be redefined past today's binary sick or not sick model, towards wellness and fitness optimization instead.

An additional trend being seen is the rise of digital fitness trackers. Fitness tracking devices are commonly worn around the wrist, neck, or on the ear, and combine specialized sensors to detect motion, steps taken, and heart rate. More advanced models can combine sensors with computing algorithms to provide a user with respiration rates, calories burned, sleep cycle analyses, and general metabolic information. Many of the fitness trackers currently on the market allow for a user to upload and share fitness data to a computer or a social network, allowing for the tracking of a user's fitness data over time.

A smartwatch refers to a mobile computing device worn on a wrist. In contrast to a mobile device that is carried by hand, such as a smartphone, a smartwatch is designed specifically for being worn on a wrist of the user. A "smartwatch" or similar device typically includes several main components, such as a head unit that includes a processor and a display, a wrist strap that allows a user to wear the smartwatch, and a charger that provides power to the smartwatch, typically when the watch is not being worn by a user.

In spite of the rising popularity of both wearable computers and fitness trackers, the wristwatch still remains a popular fashion accessory. Wristwatches can be a triumph of mechanical design, having hundreds, even thousands, of moving parts. Many luxury watches have the mechanical ability to display far more than the hours and the minutes; extra features, such as tracking eclipses or planetary motions, are termed “complications” in horology, the study of watches and clocks. Timepieces convey status and wealth, fashion and taste, and a sense of punctuality. And while many of the above mentioned wearable computers or fitness trackers seek to emulate clocks or watches on their central displays, none can replicate the mechanical intricacy or aesthetic elegance of a luxury timepiece.

SOLUTION

Briefly described, aspects of the present invention relate to a watchband with integrated electronics and a display that can provide the same functionality of a wearable computer or fitness tracker, but able to be attached to a user's desired mechanical or digital timepiece such that the timepiece's aesthetics and functionality are not impaired. Create your own hybrid smartwatch, enjoying health monitoring and notification features, simply by switching out your existing wrist straps/bands. The smart wriststrap or watchband may be seamlessly paired up with a mobile phone APP. Keep receiving your notifications through the display of this built-in fully-capable and fully-functional integrated smartwatch in a traditional watch via a smart wriststrap or watchband. A smart wearable wriststrap or watchband with its own display built in within a strap/band length in between two timepiece connection channels that receive timepiece joints for holding a standard mechanical or a digital timepiece in place. A smart wearable wriststrap or watchband with a display built in a longitudinal length of the wearable smart wriststrap or watchband and it is designed to be attachable to any traditional mechanical or digital wrist timepieces or wrist watches, adding fitness tracking, notifications and calling features to any standard wrist watch. Any standard mechanical or a digital timepiece can be used with the wearable smart wriststrap or watchband. The wearable smart wriststrap or watchband includes first and second timepiece connection channels on two free ends to receive a corresponding first and second timepiece joints for holding the standard mechanical or a digital timepiece in place by the combination of the first and second timepiece joints and the first and second timepiece connection channels.

The smart wearable wriststrap or watchband performs as a normal watchband with the electronics capability of a mobile computer and a fitness tracker. The smart wearable wriststrap or watchband has an embedded heart rate sensor, body temperature sensor, ambient temperature sensor, vibration generator, inertial sensors, and wireless communication device. The smart wearable wriststrap or watchband is powered by a rechargeable battery, which is recharged using a charging port that can be connected to a battery charger. As smartwatches take hold, many horological hardliners are sticking to their traditional watches with renewed fervour. So a smart wearable wriststrap or watchband that gives classic timepieces a smart upgrade by turning them into a hybrid timepiece with two traditional and smart types of watches into one could be just what the traditionalist has been waiting for. It's the complete smart watch with its own display built in the wriststrap or watchband that makes this smart wriststrap or watchband truly unique. You can use it for taking calls or to have it read out your text messages. The smart wriststrap or watchband has links at the ends that can be added or removed to get a better fit. The smart wearable wriststrap or watchband works with Android and iOS.

In accordance with one illustrative embodiment of the present invention, a smart wearable watchband is provided. The smart wearable watchband comprises a body having a longitudinal length including first and second free ends and a display disposed within the longitudinal length of the body. The body is configured for wearing on a wrist as a strap of a mechanical or a digital wrist timepiece and the first and second free ends of the body are removably attachable to a dial housing of the mechanical or the digital wrist timepiece.

Consistent with another embodiment, a modular smart wearable watchband is provided. The modular smart wearable watchband comprises a first smart link having a first longitudinal length including first and second free ends, wherein the first smart link including a first connector on the first free end and a first port on the second free end and wherein the first smart link including a first embedded electronic device. The modular smart wearable watchband further comprises a second smart link having a second longitudinal length including first and second free ends, wherein the second smart link including a second connector on the first free end and a second port on the second free end and wherein the second smart link including a second embedded electronic device different than the first embedded electronic device. The modular smart wearable watchband further comprises a display disposed on the first smart link, wherein the first smart link and the second smart link are together configured for wearing on a wrist as a strap of a mechanical or a digital wrist timepiece and the first free end of the first smart link and the second free end of the second smart link are removably attachable to a dial housing of the mechanical or the digital wrist timepiece.

According to yet another embodiment of the

present invention, a method of transforming a mechanical or a digital wrist timepiece into a smartwatch is provided. The method comprises providing a smart wearable watchband including a body having a longitudinal length including first and second free ends, providing a display disposed within the longitudinal length of the body, wherein the body is configured for wearing on a wrist as a strap of a mechanical or a digital wrist timepiece and the first and second free ends of the body are removably attachable to a dial housing of the mechanical or the digital wrist timepiece and converting the mechanical or the digital wrist timepiece into a hybrid watch by attaching the smart wearable watchband to the dial housing of the mechanical or the digital wrist timepiece.

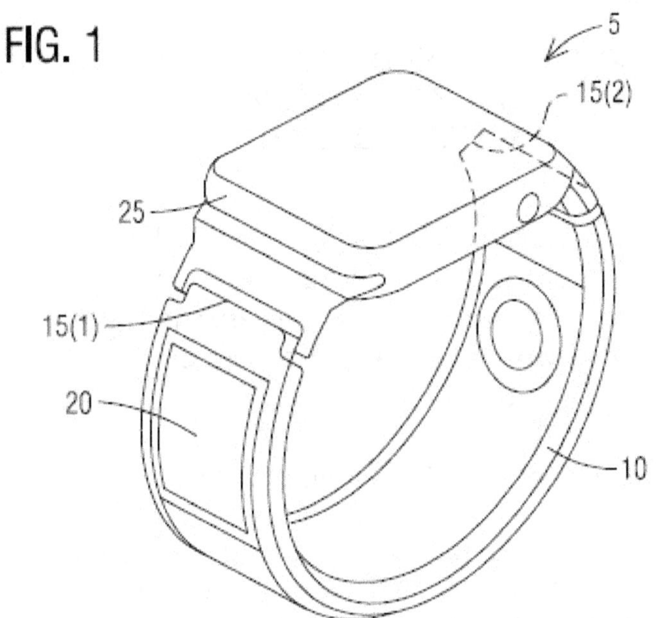
FIG. 1
5
15(2)
25
15(1)
20
10

Chapter 15

HOLLOW BATH AND HAND SOAP BAR SHELL WITH A HOLLOW INSERT OF NON-SOAP MATERIAL LOCATED WITHIN ITS CORE

Commonly provided a bath soap bar to hotel patrons generally creates unnecessary waste when it is disposed after only one or a few uses. Agreed.

My this invention (U.S. Patent Application No. 15/583280) relates generally to soap bars commonly used for human bathing or hand cleaning, and more particularly, a hollow soap bar shell with a non-soap material insert.

BACKGROUND OF PROBLEM

Soap bars are commonly used for human bathing and a variety of cleansing tasks such as hand cleaning. Unless you plan on holding up in a hotel room for an extended period of time, it's safe to say that complimentary bar of soap won't get used up. But where does that soap go after the hotel guests are gone.

In the hotel industry, it is common to provide a fresh bar of bath soap for each new hotel guest. Some luxury hotel brands also commonly provide a fresh bar of bath soap each day, even for a multi-night stay. The resulting disposal of the bath soap used the prior day creates an undesirable waste stream which must be disposed or recycled.

At least some of soap bar waste from the hotel industry goes to a company that recycles the soap to make new soap. The soap they use never reaches the landfill, helping the local environment and all of the new soap is distributed to areas of need. Hotels actually pay the recycle company to take their unused soap – $.50 per room, per month. The soap is melted down and reformed into new bars. These new soaps are packaged and sent off to charities around the world.

Bath soap however, which is commonly provided as a convenience to hotel patrons, creates unnecessary waste when it is disposed after only one or a few uses. Because the lathering and cleansing process involves removal of soap material from only the surface of the bar of soap, the interior solid core soap material is not consumed during a single or few uses, thus becomes a waste when it is no longer desirable to use the soap. This is commonly the practice when hotel patrons bath once or twice in a hotel room shower and then check out of the room and depart. The hotel housekeeping personnel discard or recycle the remaining soap bar as waste, prior to the arrival of the next guest, who expects a fresh bar of soap. This is often done to convey a quality appearance and to promote good personal hygiene.

SOLUTION

Briefly described, aspects of the present invention relate to a soap bar which minimizes resulting waste at the end of its useful life. This soap bar does not focus on subsequent recycling of soap bar remnants as a very minimal reusable structural core left to recycle. Because the lathering and cleansing process involves removal of soap material from only the surface of the bar of soap, the interior hollow core soap or non-soap material is not needed to be consumed during a single or few uses.

In accordance with one illustrative embodiment of the present invention, a bath or hand soap bar shell is provided. The bath or hand soap bar shell comprises a first soap portion formed from a first meltable soap material, a hollow core insert formed from a second material, wherein the second material is different than the first meltable soap material and a second soap portion formed from the first meltable soap material. The second soap portion is applied to the first soap portion such that the bath or hand soap bar shell fully encases the hollow core insert.

Consistent with another embodiment, a method for producing a bath or hand soap bar shell is provided. The method comprises pouring a meltable soap material into a mold cavity, forming a first soap piece in the mold cavity from the meltable soap material, positioning a hollow core insert formed from a non-soap material on the first soap piece, pouring the meltable soap material into the mold cavity to form a second soap piece and applying the second soap piece to the first soap piece as a molten soap material forming an adhesive interface therebetween when the first soap piece and the second soap piece are brought into contact with each other.

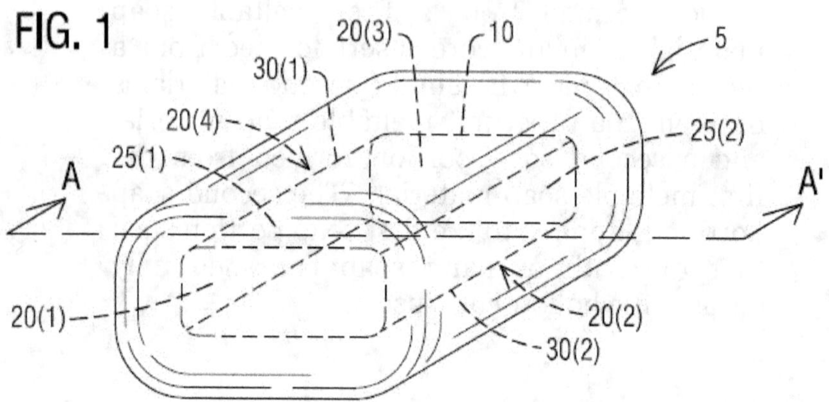
FIG. 1
20(3)
10
5
30(1)
20(4)
25(2)
25(1)
A
A'
20(1)
20(2)
30(2)

Chapter 16

CHOKING PROOF PEEL RESISTANT ADHESIVE BANDAGES PLACED ON BABIES AFTER BLOOD TESTS OR VACCINATIONS

Use of bandages on children younger than 2 years of age presents a possible choking hazard. Often, bandages are placed on the fingers and hands of young children after routine blood tests or vaccinations. Inevitably, the bandaged hand or finger goes into the child's mouth, creating the potential to choke. Choking is a common cause of injury and death in children younger than 5 years. We can change this predicament. Agreed.

My this invention (U.S. Patent Application No. 15/601242) relates generally to adhesive bandages commonly used for babies, and more particularly, choking proof and peel resistant adhesive bandages which form a ring-like grip that keeps the choking proof adhesive bandage in place despite any attempt to take it off by a baby.

BACKGROUND OF PROBLEM

A breathing emergency is any respiratory problem that can threaten a person's life and choking is an example of breathing emergencies. Choking is a common breathing emergency. It occurs when the person's airway is partially or completely blocked. If a conscious person is choking, his or her airway has been blocked by a foreign object, such as a piece of food or a small toy; by swelling in the mouth or throat; or by fluids, such as vomit or blood. With a partially blocked airway, the person usually can breathe with some trouble. A person with a partially blocked airway may be able to get enough air in and out of the lungs to cough or to make wheezing sounds. The person also may get enough air to speak. A person whose airway is completely blocked cannot cough, speak, cry or breathe at all.

Choking is a common cause of injury and death in children younger than 5 years. Because young children put nearly everything in their mouths, small, non-food items, such as safety pins, small parts from toys and coins, often cause choking. However, food is responsible for most of the choking incidents in children. Since choking remains a significant danger to children younger than 5 years, the American Academy of Pediatrics (AAP) further recommends keeping the following foods, and other items meant to be chewed or swallowed, away from young children.

For prevention, do not give young children round, firm foods such as hot dogs and carrot sticks unless chopped into pieces ½ inch or smaller. Keep small objects such as safety pins, small parts from toys and coins away from small children. Make sure that toys are too large to be swallowed. Make sure that toys have no small parts that could be pulled off. If you are unsure whether an object is safe for young children, test it by trying to pass it through a toilet paper roll. If it fits through the 1¾-inch diameter roll, it is not safe for young children.

Although food items cause most of the choking injuries in children, toys and household items also can be hazardous. Balloons, when broken or un-inflated, can choke or suffocate young children who try to swallow them. According to the Consumer Product Safety Commission (CPSC), more children have suffocated on non-inflated balloons and pieces of broken balloons than any other type of toy.

Signals of choking include: coughing, either forcefully or weakly, clutching the throat with one or both hands, inability to cough, speak, cry or breathe, making high-pitched noises while inhaling or noisy breathing, panic, bluish skin color, and losing consciousness if blockage is not removed.

A bandage is a piece of material used either to support a medical device such as a dressing or splint, or on its own to provide support to or to restrict the movement of a part of the body. Bandages pose a choking threat. Use of bandages on children younger than 2 years of age presents a possible choking hazard. Often, bandages are placed on the fingers and hands of young children after routine blood tests or vaccinations. Inevitably, the bandaged hand or finger goes into the child's mouth, creating the potential to choke. Doctors always caution parents to closely supervise children who are wearing such bandages, and tell them to remove bandages promptly once bleeding has stopped.

SOLUTION

Briefly described, aspects of the present invention relate to choking proof and peel resistant bandages with a combination of a standard bandage and two transverse longer bandage strips which form a ring that keeps the choking proof bandage in place even if it is tampered with by a baby to take it off from their skin. By being impossible to be removed from the skin such as a hand, a finger or an arm due to extra bandage area of adhesion available from the two transverse longer bandage strips the choking proof bandage almost eliminates the potential to choke in babies.

In accordance with one illustrative embodiment of the present invention, an adhesive bandage is provided. The adhesive bandage includes a first and a second bandage arms and a third and a fourth bandage arms. The first and second bandage arms are transverse to the third and fourth bandage arms. The first and second bandage arms are longer in length to the third and fourth bandage arms. The adhesive bandage further includes a medicated pad at an intersection of the first and second bandage arms and the third and fourth bandage arms.

Consistent with another embodiment, an adhesive bandage is provided. The adhesive bandage includes a bandage strip with a channel in a middle portion to thread through an object. The adhesive bandage further includes a clip threaded through the channel of the bandage strip. The clip is configured to be placed on an arm or a thigh of a baby. The adhesive bandage further includes a medicated pad at an intersection of the middle portion of the bandage strip and the clip.

Consistent with yet another embodiment, an adhesive bandage is provided. The adhesive bandage includes a bandage strip with a channel in a middle portion to thread through an object. The adhesive bandage further includes a Velcro band threaded through the channel of the bandage strip. The Velcro band is configured to be placed on an arm or a thigh of a baby. The adhesive bandage further includes a medicated pad at an intersection of the middle portion of the bandage strip and the Velcro band.

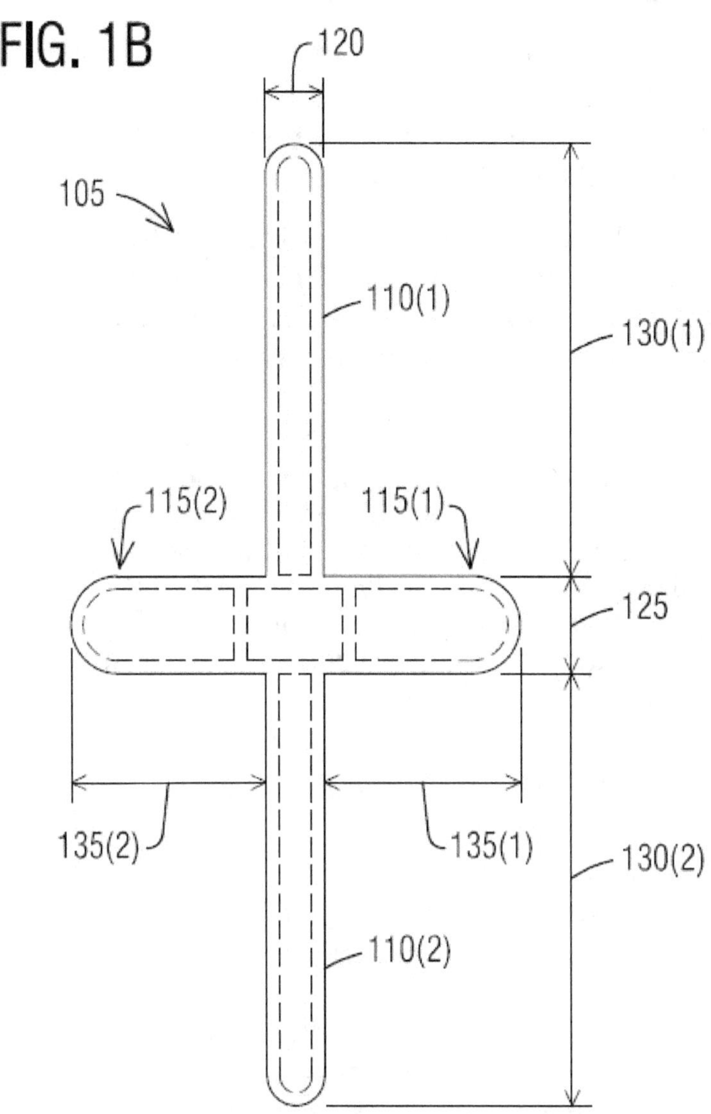
FIG. 1B
120
105
110(1)
130(1)
115(2)
115(1)
125
135(2)
135(1)
130(2)
110(2)

Chapter 17

PACKAGING MULTI-PACK OF CONTAINERS HAVING COMMON GEOMETRIC SHAPE WITH SHEET OR BAND PACKAGING DEVICE

We've all seen images of wildlife killed or maimed by plastic six-pack rings. We need to eliminate this particular threat to our shared environment. Agreed.

My this invention (U.S. Patent Application No. 15/690191) relates generally to a packaging device for packaging a multi-pack of containers, and more particularly relate to a sheet or a band packaging device for packaging a multi-pack of cans or bottles having a common geometric shape.

BACKGROUND OF PROBLEM

We've all seen images of wildlife killed or maimed by plastic six-pack rings. We've seen these things scattered across the landscape and floating in our waters. The good news is that each one of us has the power to eliminate this particular threat to our shared environment.

Six-pack rings, also known as "yokes", are a common and persistent environmental problem that needs to be resolved. Although only 50 years old, this product can be found nearly everywhere and causes a substantial amount of environmental damage. Although we refer to them as 6-pack rings, this product can actually have any number of rings. And though they are commonly used as packaging for aluminum drink cans, 6-pack rings can be found on many types of products.

The best solution is to not create the problem in the first place. If consumer demand for products packaged in 6-pack rings declined, manufacturers would stop producing them. That's how our economy works. As a consumer, our purchasing decisions have power. Looking for products that are in paper packaging or buying individual products and carrying them home in a reusable bag is one solution of this problem.

If one feels extra motivated, one could write to companies that use this packaging for their products. Let them know how unhappy one is with this decision and that you can boycott their products that include 6-pack rings. Consumer opinion does count and has changed how industries design and manufacture products.

Cutting every hole (not just the primary rings) greatly reduces the chance that an animal will get entangled in the product. This is critical even if you plan on recycling the rings. The rings could blow off a truck or out of the recycling plant and get into the environment. A few seconds of your time could save a life (or several lives). However, how many consumers would regularly do this cutting every time they throw or recycle six-pack rings.

Six-pack rings are made from low-density polyethylene (LDPE). This is the same material that is used to make plastic bags and many types of plastic containers. While the rings themselves are not usually marked, LDPE is identified with the recycling code 4 and is completely recyclable. The code is just an arbitrary number and is used to identify the type of plastic used in a product.

Recycling plastic products has several positive effects. First, it keeps the item from taking up space in a landfill. Second, recycling removes the potential for the product to get out into the environment. Third, recycling used plastic produces less air and water pollution than the production of virgin plastic. Fourth, using recycled plastic reduces our dependence on petroleum. Finally, recycling closes the resource loop. For true sustainability, all waste products need to become feed stock for new products.

Products made from LDPE can't usually be recycled back into their original form. The recycling process changes the chemical structure of the plastic. However, LDPE products can be recycled into more durable products such as trash cans, lumber, and floor tile.

Six-pack beverage can rings are not the largest or most dangerous environmental threat we face. But they do create a tangible and substantial hazard that cannot and should not be ignored. Especially since they are an extremely unnecessary product.

SOLUTION

Briefly described, aspects of the present invention relate to a multi-pack package device including a carrier that includes a plurality of apertures for receiving multiple containers such that each of the plurality of apertures being surrounded in two dimensions by a sheet or a band. The multi-pack package device further includes a plurality of bonds formed by bonding a neck surface of a container of the multiple containers to the sheet or the band and a plurality of perforations-based structures including perforations arranged to create a tear line that passes through a particular aperture of the plurality of apertures at a location where the neck surface is bonded to the sheet or the band.

In accordance with one illustrative embodiment of the present invention, a multi-pack packaging device is provided for carrying multiple cans or bottles each having a common geometric shape and adapted to permit release of each of the multiple cans or bottles individually. The multi-pack packaging device comprises a carrier, formed from a sheet of a paper-based material or a band of a resilient material, that includes a plurality of apertures for receiving the multiple cans or bottles, each of the plurality of apertures being surrounded in two dimensions by the paper-based material or the resilient material as the sheet of the paper-based material or the band of resilient material extends about at least a portion of a periphery of each of the multiple cans or bottles while forming a closed loop about the multiple cans or bottles. The multi-pack packaging device further comprises a plurality of bonds formed by bonding a neck surface of a can or a bottle of the multiple cans or bottles to the sheet of the paper-based material or the band of resilient material. Each of the plurality of bonds associated with a different one of the plurality of apertures and configured to hold one can or bottle of the multiple cans or bottles. The multi-pack packaging device further comprises a plurality of perforations-based structures including perforations arranged to create a tear line that passes through a particular aperture of the plurality of apertures at a location where the neck surface is bonded to the sheet of the paper-based material or the band of resilient material. The perforations are configured to enable the plurality of perforations-based structures to be torn from the multi-pack packaging device to cause the particular aperture to no longer be surrounded in two dimensions by the paper-based material or the resilient material, and to enable a can or a bottle of the multiple cans or bottles to be removed from the multi-pack packaging device.

Consistent with another embodiment, a multi-pack packaging device comprises multiple containers. The multi-pack packaging device comprises a carrier that includes a plurality of apertures for receiving the multiple containers. Each of the plurality of apertures being surrounded in two dimensions by a sheet or a band. The multi-pack packaging device further comprises a plurality of bonds formed by bonding a neck surface of a container of the multiple containers to the sheet or the band. Each of the plurality of bonds associated with a different one of the plurality of apertures and configured to hold one container of the multiple containers. The multi-pack packaging device further comprises a plurality of perforations-based structures including perforations arranged to create a tear line that passes through a particular aperture of the plurality of apertures at a location where the neck surface is bonded to the sheet or the band. The perforations are configured to enable the plurality of perforations-based structures to be torn from the multi-pack packaging device to cause the particular aperture to no longer be a loop, and to enable a container of the multiple containers to be removed from the multi-pack packaging device.

Consistent with yet another embodiment, a method of packaging a multi-pack of containers is provided. The method comprises providing multiple containers, providing a carrier that includes a plurality of apertures for receiving the multiple containers, each of the plurality of apertures being surrounded in two dimensions by a sheet or a band, providing a plurality of bonds formed by bonding a neck surface of a container of the multiple containers to the sheet or the band, each of the plurality of bonds associated with a different one of the plurality of apertures and configured to hold one container of the multiple containers and providing a plurality of perforations-based structures including perforations arranged to create a tear line that passes through a particular aperture of the plurality of apertures at a location where the neck surface is bonded to the sheet or the band, the perforations are configured to enable the plurality of perforations-based structures to be torn from the multi-pack packaging device to cause the particular aperture to no longer be a loop, and to enable a container of the multiple containers to be removed from the multi-pack packaging device.

3/4

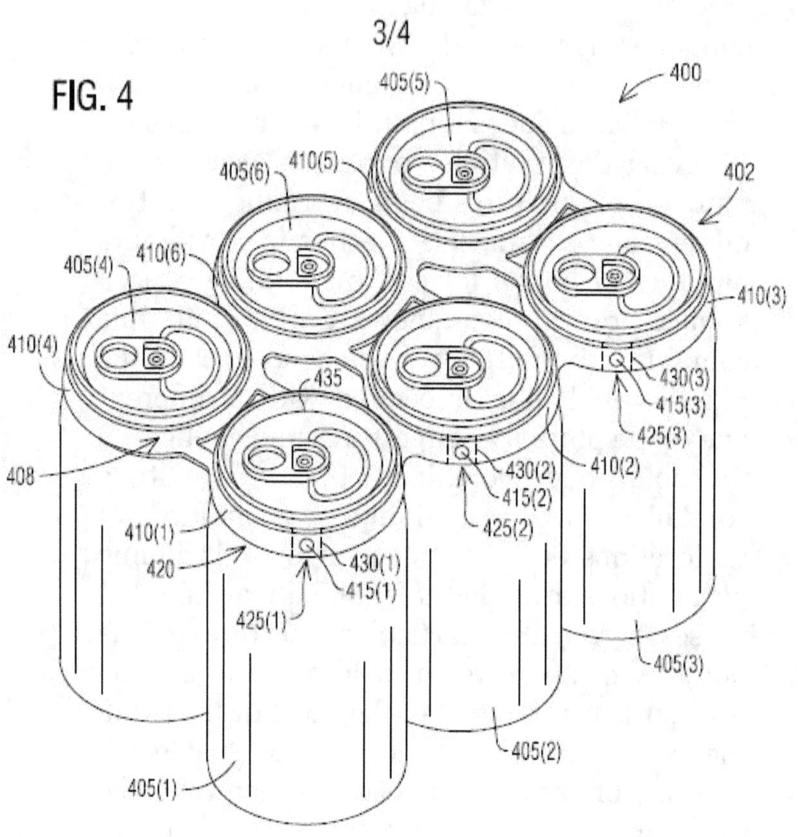
FIG. 4
400
405(5)
410(5)
405(6)
402
405(4)
410(6)
410(3)
410(4)
435
430(3)
415(3)
425(3)
408
430(2)
410(2)
415(2)
425(2)
410(1)
420
430(1)
415(1)
425(1)
405(3)
405(2)
405(1)

Chapter 18

PROTECTING FROM AN UNAUTHORIZED REMOVAL OR THEFT OF REGISTRATION STICKER TAG FROM LICENSE PLATE OF VEHICLE

Unethical individuals go around stealing your license plate registration tag stickers from parked vehicles. This is a problem because replacing them costs money and time. Agreed.

My this invention (U.S. Patent Application No. 15/787108) relates generally to means for protecting against an unauthorized removal or theft of a registration sticker tag from a license plate of a vehicle, and more particularly relate to a kit including an auto license plate frame, a protection frame insert and/or a transparent cover for protection of a registration sticker tag.

BACKGROUND OF PROBLEM

License plate registration stickers come at a high cost, especially in some states, which encourages some unethical individuals to go around stealing them instead. It's an ongoing problem-- crooks stealing license plate tags and putting them on unregistered cars.

You can deter someone from stealing the registration sticker on your license plate by cross cutting it with a razor blade. If they try to pull it off, it will only come off in pieces. Of course, the thief could always pull both pieces of the sticker but removing pieces makes it much more difficult. This trick will still not work even if you've prevented them from removing the whole sticker. You may have to worry about getting a replacement. It is a big hassle when a thief steals a registration sticker tag off a car. One has to wait for hours at the Department of Motor Vehicle (DMV) to get a new registration sticker tag.

There's a popular product to protect you, but if you use it, you could wind up in trouble. Here's the twist. These products are sold all over the state -- covers that protect your license plates from theft or damage. It's perfectly legal to sell them, but don't try to use one. A license plate protector -- a clear plastic shield can be fastened over a license plate. Now it would be pretty hard to steal the registration sticker tags, except there is just one little problem. "It's illegal to have on your car," in California state. It's illegal in California to use these covers, even though they're sold all over the state. The law says you can't have anything that completely covers the license plate. Vehicle Code Section 5201 says "No portion of a cover shall rest over the license plate number." Not even if you can see right through it. You could be stopped by police and you could receive a ticket.

You can buy license plate frames that partially cover only the registration sticker tags so it's harder to peel them off. Those frames are legal if they don't cover the license numbers. However, partially covering the registration sticker tags does not deter the thieves from stealing them. Likewise, to deter tag thieves one can slice the registration sticker tag with a razor so that if it comes off, it doesn't come off in one piece. However, when removed even partially means one has to get a new one issued.

SOLUTION

Briefly described, aspects of the present invention relate to kits including an auto license plate frame, a protection frame insert and/or a transparent cover for preventing an unauthorized removal or theft of a registration sticker tag from a license plate of a vehicle.

In accordance with one illustrative embodiment of the present invention, a license plate frame kit for use with a vehicle and a license plate having apertures for receiving bolts comprises an auto license plate frame having a first plate with a first perimeter. The first plate includes a plurality of first through holes aligned with the apertures in the license plate and at least one of a protection frame insert and a transparent cover. The protection frame insert may be configured for protection of a registration sticker tag. The protection frame insert has a second plate with a second perimeter and a second through hole configured for alignment with at least one first through hole of the plurality of first through holes of the first plate such that a screw may pass through the aligned the at least one first through hole and the second through hole to attach the auto license plate frame and the protection frame insert to a license plate mount. The transparent cover may be positioned between the auto license plate frame and the license plate. The transparent cover just covers only the registration sticker tag for preventing removal of the registration sticker tag displayed on the license plate.

Consistent with another embodiment, a universal-reconfigurable license plate frame kit for use with a vehicle and a license plate having apertures for receiving bolts is provided. The kit comprises an auto license plate frame having a first plate with a first perimeter and a first protection frame insert. The first plate includes a plurality of first through holes aligned with the apertures in the license plate. The first plate has a back side facing a license plate mount on the vehicle and a first corner of the first perimeter. The back side having a first frame retention tab and a second frame retention tab proximate the first corner of the first perimeter of the first plate. The first protection frame insert configured for protection of a registration sticker tag. The first protection frame insert has a second plate with a second perimeter and a second through hole configured for alignment with at least one first through hole of the plurality of first through holes of the first plate. The first protection frame insert is configured to slide through a guide defined by the first frame retention tab and the second frame retention tab.

Consistent with yet another embodiment, a first protection frame insert comprises a plate with a perimeter configured for protection of a registration sticker tag. The plate has a corner defined in the perimeter. The corner of the plate having a first through hole configured for alignment with at least one second through hole of a plurality of second through holes of an auto license plate frame that is configured for use with a vehicle and a license plate having apertures for receiving bolts such that a screw may pass through the aligned the first through hole and the at least one second through hole to attach the first protection frame insert and the auto license plate frame to a license plate mount on the vehicle.

FIG. 1

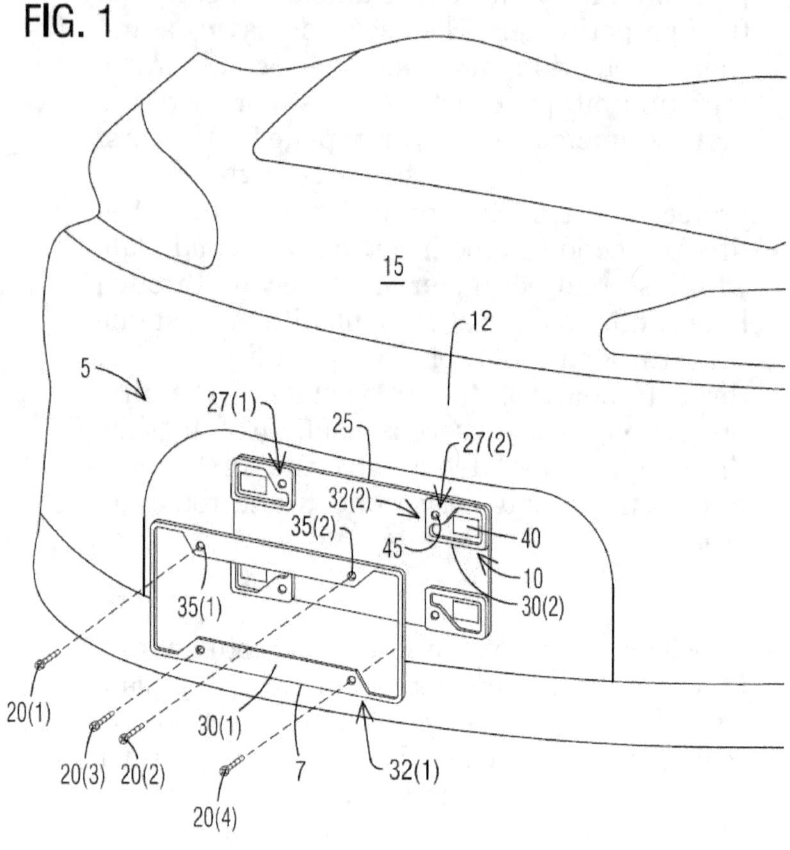

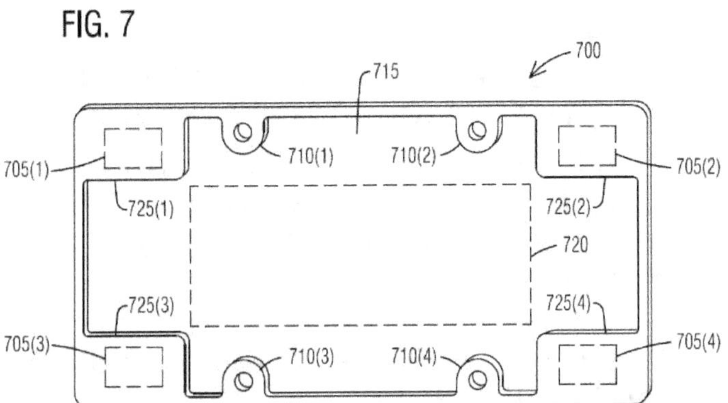
FIG. 7
700
715
705(1)
710(1)
710(2)
705(2)
725(1)
725(2)
720
725(3)
725(4)
705(3)
710(3)
710(4)
705(4)

Chapter 19

AUDIO APP USER INTERFACE FOR PLAYING AN AUDIO FILE OF A BOOK THAT HAS ASSOCIATED IMAGES CAPABLE OF RENDERING AT APPROPRIATE TIMINGS IN THE AUDIO FILE

There are no means for showing actual images such as figures, drawings, graphs of a hard copy book or eBook at the time of audio playback of an audio file of an audio book. This feature is needed. Agreed.

My this invention (U.S. Patent Application No. 15/801565) relates generally to an audio application (APP) user interface that is configured to display images included in a book while playing an audio file of the book, and more particularly relate to rendering electronic images such as drawings or figures of graphs appearing a paper or hard copy book in a user interface display along with an audio recitation of the book from the audio file of the narrated book.

BACKGROUND OF PROBLEM

Publishers and/or authors frequently offer audio versions of their books or other written works to consumers. Audio books and other narration audio recordings are often sold on a recorded medium, such as a compact disc or cassette tape. Audio recordings, such as audio books, are often offered in formats and/or packaging.

Publishers and/or authors frequently offer audio versions of their books or other written works to consumers. A user will often listen to an audio book or other narration audio (such as spoken word recordings of magazine or newspaper articles, podcasts, text-to-speech audio content, etc.) using a device that includes both the capability to play audio content and the capability to display visual content. For example, a portable computing device, mobile phone, tablet computer, or electronic book reader ("e-book reader") capable of playing audio content may also include a display device, such as a touch screen display. Although many such computing devices are frequently utilized by users to listen to digital audio books and other narrated audio recordings, the associated display capabilities of the devices are often underutilized during playback of the audio content. For example, a user may simply be presented with basic audio controls (such as play and pause options) on the display device during playback of the audio content.

SOLUTION

Briefly described, aspects of the present invention relate to an audio file of an audio book of a hard copy book or eBook that shows actual images such as figures, drawings, graphs of a hard copy book or eBook at the time of audio playback in a synchronized manner so all available or assorted visual information present in a book is offered for viewing in a user interface of a user device.

In accordance with one illustrative embodiment of the present invention, a system comprises a data store that stores an audio base content of text present in a hard copy book or an eBook having a plurality of pages with one or more actual book images and a book images base content present in the one or more actual book images of the plurality of pages of the hard copy book or the eBook. The book images base content has content synchronization information that correlates positions of the one or more actual book images within the audio base content with a physical location of appearance of each actual book image of the one or more actual book images on a particular page of the hard copy book or the eBook. The system further comprises a computing device, comprising one or more processors, in communication with the data store and that is configured to at least: receive a request for playback of the audio base content stored in the data store and in response to the request for playback of the audio base content, retrieve or stream from the data store in a sequence the one or more actual book images within a user interface of an audio application (APP) so that each actual book image appears in the user interface at a designated time in an audio stream of the hard copy book or the eBook. The designated time is based on the content synchronization information. Consistent with another embodiment, a computer-implemented method for facilitating presentation of a plurality of actual book images is provided. The computer-implemented method comprises under control of one or more computing devices configured with specific computer executable instructions, receiving for storage in a data store an audio base content of text present in a hard copy book or an eBook having a plurality of pages with one or more actual book images, receiving for storage in the data store a book images base content present in the one or more actual book images of the plurality of pages of the hard copy book or the eBook, the book images base content having content synchronization information that correlates positions of the one or more actual book images within the audio base content with a physical location of appearance of each actual book image of the one or more actual book images on a particular page of the hard copy book or the eBook, receiving a request for playback of the audio base content stored in the data store, and in response to the request for playback of the audio base content retrieving or streaming from the data store in a sequence the one or more actual book images within a user interface of an audio application (APP) so that each actual book image appears in the user interface at a designated time in an audio stream of the hard copy book or the eBook, wherein the designated time is based on the content synchronization information.

Consistent with yet another embodiment, a system comprises a data store that stores: an audio base content of a hard copy book or an eBook having a plurality of pages with one or more actual book images in an audio file, a book images base content present in the one or more actual book images of the plurality of pages of the hard copy book or the eBook in a video file associated with the audio file, information regarding the one or more actual book images, wherein each actual book image of the one or more actual book images comprises a video content associated with at least one subject referenced by the audio base content and providing additional information not included within the audio base content regarding the at least one subject, content synchronization information that correlates positions within the audio base content of the audio file with the one or more actual book images of the video file. The system further comprises a computing device, comprising one or more processors, in communication with the data store and that is configured to at least: receive a request for playback of the audio file stored in the data store, in response to the request for playback of the audio file: retrieve, from the data store, the one or more actual book images associated with the audio base content, the one or more actual book images including visual images representing figures, drawings or graphs, receive, from the data store, a first actual book image associated with a first audio content portion and a second actual book image associated with a second audio content portion, wherein the first actual book image is different than the second actual book image, and automatically present for display at least the first actual book image and the second actual book image at different times during playback of corresponding portions of the audio base content, such that (a) the first actual book image is presented for display during playback of the first audio content portion corresponding to a first page of the hard copy book or the eBook and (b) the second actual book image is presented for display during playback of the second audio content portion corresponding to a second page of the hard copy book or the eBook.

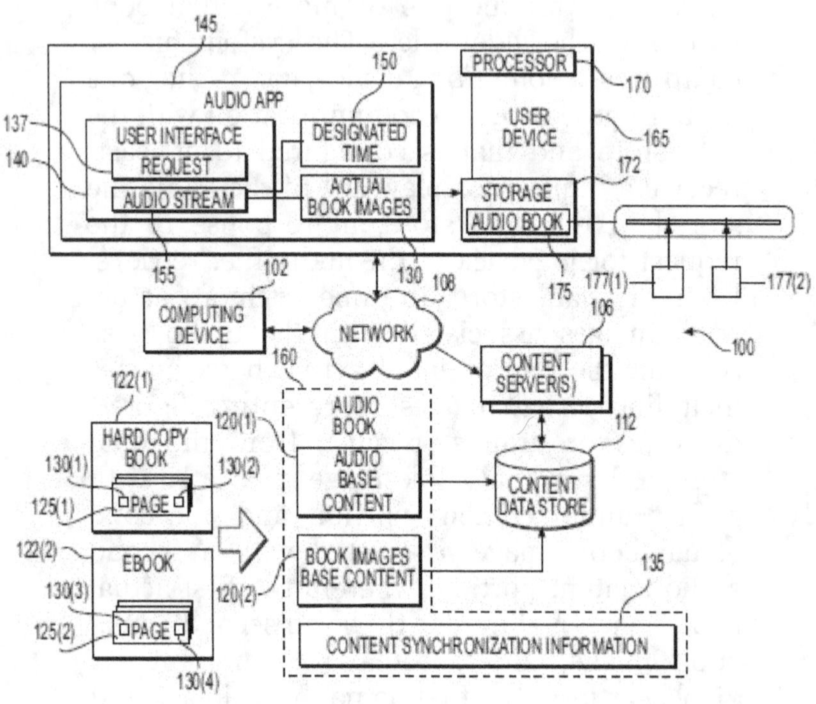
145
150
PROCESSOR
170
AUDIO APP
137
USER INTERFACE
DESIGNATED TIME
USER DEVICE
165
140
REQUEST
172
AUDIO STREAM
ACTUAL BOOK IMAGES
STORAGE
AUDIO BOOK
155
102
130
108
175
177(1)
177(2)
COMPUTING DEVICE
NETWORK
106
100
160
CONTENT SERVER(S)
122(1)
120(1)
AUDIO BOOK
112
HARD COPY BOOK
AUDIO BASE CONTENT
130(1)
130(2)
CONTENT DATA STORE
125(1)
PAGE
122(2)
EBOOK
BOOK IMAGES BASE CONTENT
135
130(3)
120(2)
125(2)
PAGE
CONTENT SYNCHRONIZATION INFORMATION
130(4)

Chapter 20

A CASE STUDY FIT TO PUBLISH: HOW THE USE OF DUALITY INNOVATION PRINCIPLE CAN INCREASE SALES FOR THE COMPANY WE KNOW AS "CROCS"

CROCS makes at least 4 lines of products named – clogs, flip flops, sandals and shoes. They have about 25 models of clogs ALL with round toe cap for both men and women. They have about 25 models of flip flop. They ALL consist of a flat sole held loosely on the foot by a Y-shaped strap known as an upper that passes between the first and second toes and around both sides of the foot. The sandals were mostly SINGLE colored or were multi-colored but poorly executed. ALL flat sole shoes were round or oval ONLY. These features need to be expanded to offer MORE design choices as different designs speak to different people in different ways (resulting in increased sales). Agreed.

CLOG dual design

Offering 25 clogs of single round toe cap design to a customer who likes this design cannot result in significantly more sales. But changing the design somewhat can appeal to other customers who don't like the round head design. Yes, one can come to this conclusion on its own by being more thorough and diligent in finding ways to increase sales. But if you apply the duality innovation principle then you could assign the round toe cap design a logic "1". Then the question is what is the design for a logic "0". My thought is that this logic "0" design is an open toe rather than a covered toe. The logic "0" design could mean a different design to someone else. I picked the open toe design because the logic "1" is hidden toe so logic "0" would be visible toe. The other reason supporting my choice is that women like to show off their pedicure treatment of the toenails.

FLIP FLOP dual design

All flip flop designs based on a Y-shaped strap to a customer may look the same and can make happy only those who prefer this configuration. This is logic "1" design. See a Y-shaped strap in FIG. 1 below.

What would really increase sales if we can bring more people under the flip flop tent. Now this plan can be executed by experimenting with other possible known or new designs. New designs are hard to master and may not become popular. But one can draw from other cultures or countries to pick something which will work in USA.

One design I preferred is logic "0" design that has a loop for the toe and a strap over the foot. Some folks who do not like the Y-shaped strap design may opt for this different concept. See loop design in FIG. 2 below.

SANDALS dual design

Many sandals of the crocs website were single colored and some were dual colors but did not make good color look. I am not expert on color preferences by males or females but can see why some combination of colors may be more attractive. One good possibility is making the sole top layer of a light color and the sole layer base may be made a dark color. The top straps can vary based on the choice of these two colors. The key idea is that a smart choice of color combinations may work and sell more than others. I am sure Crocs already have this know-how but perhaps lacks in execution.

Shoes sole dual design

All shoe soles on the crocs website were in a round or oval shape. I call this logic "0". See FIG. 3 below.

What is needed is one other option which is different and gives some trail blazers a choice which is unique and makes a statement. The design choice is a square/rectangular shape for the sole of a shoe/clog/flip flop/sandal. This shape would work well with flip flops. See FIG. 4 below.

Chapter 21

MORE ON THE WEBSITE

Refer to the website:

www.dualityinnovation.com

FOR FEEDBACK:

MY EMAIL ADDRESS IS:

singh_sanjeevk@yahoo.com

www.ingramcontent.com/pod-product-compliance
Lightning Source LLC
Chambersburg PA
CBHW052318220526
45472CB00001B/171